Plant Physiological Aspects of Phenolic Compounds

Edited by Marcos Soto-Hernández,
Rosario García-Mateos
and Mariana Palma-Tenango

Published in London, United Kingdom

IntechOpen

Supporting open minds since 2005

Plant Physiological Aspects of Phenolic Compounds
http://dx.doi.org/10.5772/intechopen.77494
Edited by Marcos Soto-Hernández, Rosario García-Mateos and Mariana Palma-Tenango

Contributors
Shadab Gharaati Jahromi, Norma Francenia Santos-Sánchez, Raúl Salas-Coronado, Claudia Villanueva-Cañongo, Beatriz Hernández-Carlos, Muhammad Bilal Hussain, Sadia Hassan, Marwa Waheed, Muhammad Adil Farooq, Ahsan Javed, Ali Tahir, Mariana Sefora Sousa, Diego Buarque, Jovan Lara Jr., Andres Cruz-Hernández, Marcela Vargas-Hernández, Gerardo Vázquez-Marrufo, Carlos Agustín Aguilar-Ruiz, Marco Antonio González-Márquez, Oscar Rocha, Diego Roberto Cerna-Pantoja, Stefania D'Angelo

First published in London, United Kingdom, 2019 by IntechOpen
IntechOpen is the global imprint of INTECHOPEN LIMITED, registered in England and Wales, registration number: 11086078, The Shard, 25th floor, 32 London Bridge Street
London, SE19SG - United Kingdom
Printed in Croatia

British Library Cataloguing-in-Publication Data
A catalogue record for this book is available from the British Library

Additional hard and PDF copies can be obtained from orders@intechopen.com

Plant Physiological Aspects of Phenolic Compounds
Edited by Marcos Soto-Hernández, Rosario García-Mateos and Mariana Palma-Tenango
p. cm.
Print ISBN 978-1-78984-033-9
Online ISBN 978-1-78984-034-6
eBook (PDF) ISBN 978-1-78985-640-8

For EU product safety concerns:
IN TECH d.o.o., Prolaz Marije Krucifikse Kozulić 3, 51000 Rijeka, Croatia, info@intechopen.com or visit our website at intechopen.com.

Meet the editors

Dr. Marcos Soto Hernández is a pharmacist from the National University of Mexico and completed his PhD at the University of Wales, Cardiff, UK. He is now a full professor at Colegio de Postgraduados where he conducts research in phytochemistry and bioactive natural products. He has established collaboration with research groups in the UK, the Netherlands, Spain, and other groups in Mexico. He has received several awards locally and abroad. At present, his main line of research is related to bioguided isolation of secondary metabolites with importance in medicine and agriculture. In addition, the potential of local aromatic plants is part of his recent research.

Dr. Mariana Palma-Tenango is an engineer agronomist from the Universidad Autónoma Chapingo and holds a PhD in Plant Physiology from Colegio de Postgraduados, Mexico. She has performed teaching duties at the National Autonomous University of Mexico. She is a postdoctoral researcher at ESIME-Instituto Politécnico Nacional, a reviewer of several journals, and an editor. She has participated in the organization of meetings and symposiums in Mexico and is a supervisor of master's and PhD students. Her research line is phytochemistry and medicinal and aromatic plants.

Dr. Ma. Rosario García-Mateos is a chemist at the University of Puebla, Mexico, and holds a PhD degree from Colegio de Postgraduados in Plant Physiology. Her research line is related to the nutraceutical properties and biological activity of horticultural crops. She has supervised master's and PhD theses, has published more than 89 papers, and has been a reviewer of several scientific journals. She has established collaboration with many national and international groups.

Contents

Preface

The phenolic compounds formed by the shikimic pathway constitute one of the most numerous groups of secondary metabolites described in plants. These phytochemicals perform a variety of functions, such as defense against pathogens and herbivores, signaling mechanisms, fruit and flower pigments, and protection against ultraviolet light. They are components of plant cells and contribute to the adaptation of plants in the environment. During the evolution of vascular plants, they were the first to be synthesized (lignins) and mark the differences between the structural diversity of phenolic compounds among plants and their genetic variability. Several biotic and abiotic factors are also part of this structural diversity. Research on these compounds has increased due to the number of molecules that are involved and the different biological activities that are observed, such as antioxidant activity or their physiological function in plants. It is important to know the methods of extracting molecules, the biosynthesis routes, and their relationship with the activities that can benefit from their consumption. Therefore, the book includes chapters that provide information on extraction and optimization techniques, biosynthetic pathways, and the identification and characterization of miRNAs involved in the regulation of their biosynthesis.

Dr. Marcos Soto-Hernández
Full Professor,
Department of Botany,
Postgraduate College,
Mexico

Dra. Mariana Palma-Tenango
Postdoctoral reseacher,
Postgraduate in Systems,
National Polytechnic Institute,
Mexico

Dra. María del Rosario García-Mateos
Full Professor,
Department of Plant Science,
Institute of Horticulture,
Autonomous University Chapingo,
Mexico

Section 1

Extraction of Phenolic Compounds

Chapter 1

Extraction Techniques of Phenolic Compounds from Plants

Shadab Gharaati Jahromi

Abstract

Phenolic derivatives are one of the most important compounds that were found in secondary metabolites in plants. According to their various applications in agriculture, food, chemical and pharmaceutical industries, interests in reviewing different procedures of extraction of these compounds from plants have increased. In this chapter, we would like to have an overview on the extraction procedures that have been used in isolating phenolic compounds from plants until this time, including liquid-liquid extraction (LLE), ultrasound-assisted extraction (UAE), microwave-assisted extraction (MAE) and supercritical fluid extraction (SFE). In the following, advantages and disadvantages of these techniques and methods will be discussed and explained. In addition, in the last part of this chapter, various methods for purification and identification of phenolic compounds will be presented.

Keywords: plants, phenolic compounds, bioactive compounds, extraction techniques, identification techniques

1. Introduction

Plants are rich and valuable resources of bioactive phenolic. They can be utilized in various fields such as antioxidant, antimicrobial, anti-inflammatory, antitumor, antiviral, analgesic and antipyretic [1, 2]. Hence, they have attracted the attention of many health professionals and many organizations and health care systems increasingly recommend the daily consumption of fruits and vegetables [3, 4].

Phenolics are one of the major and diverse group of active compounds in the plants which have at least one aromatic ring and one or more hydroxyl groups in their structures [5, 6]. They can be divided in two categories. First category consists of soluble compounds such as flavonoids, quinones, phenylpropanoids which can be discovered in plant cell vacuole and the second category consists of insoluble compounds such as lignins, condensed tannins and hydroxyl cinnamic acid which are detected in cell-wall bound [7]. All of these groups are involved in many processes in plants. Various classifications of phenolic compounds based on various subgroups and structures have been listed in **Table 1**. Due to importance and worthiness of phenolic compounds on human health, many researches have been so far done in order to synthesizing various compounds like natural compounds using synthetic sequence procedures [8–11].

Based on various structures of phenolics, they have different physical and chemical characteristics which are very significant and emphasize on the extraction

Basic group	Subgroup	Structure
Phenolic acids	Hydroxybenzoic acid derivatives	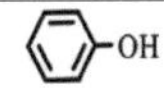 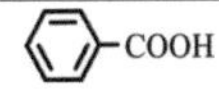
	Hydroxycinnamic acid derivatives	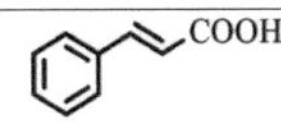 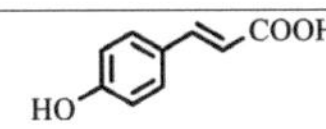
Flavonoids	Flavones	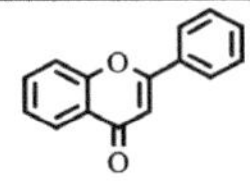
	Flavanones	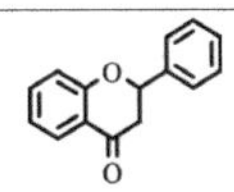
	Flavonols	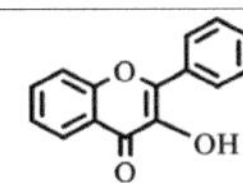
	Isoflavones	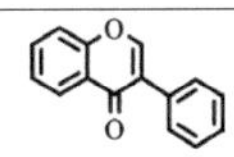
	Aurones	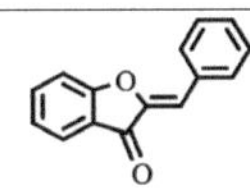
	Chalcones	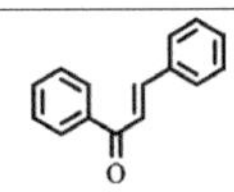
	Anthocyanins	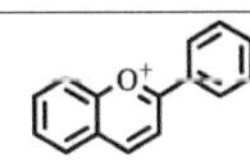
Tannins	Hydrolysable tannin	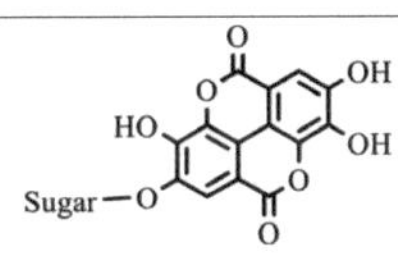
	Condensed tannins	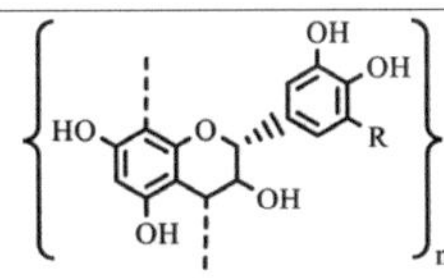
Miscellaneous group	Lignans	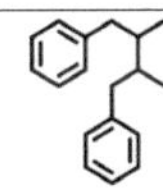
	Resveratrol	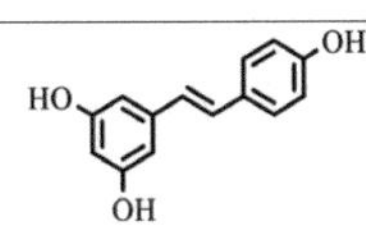
	Coumarins	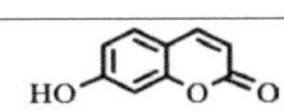
	Lignins	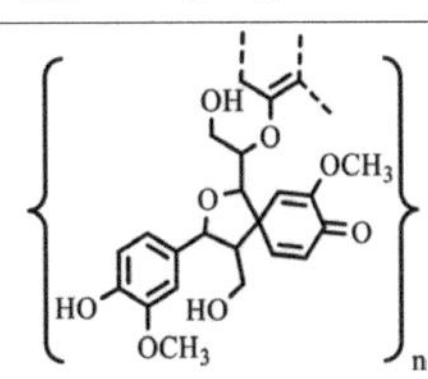

Table 1.
Classification of phenolic compounds based on various subgroups and structures.

processes [12]. Therefore having knowledge about various methods of extraction, identification and quantification of phenolic compounds are essential and useful. In this chapter various techniques for extractions of phenolic compounds with their advantage and disadvantage will be investigated. Finally in the last section of this chapter, various methods have been discussed to identify and evaluate their quality.

2. Liquid-liquid extraction (LLE)

Scientists using various solvents have studied and investigated the extraction of phenolic compounds of different parts of plants such as leaves and seeds. Based on simplicity and cheapness of this extraction method, various polarities of solvents and under different temperature and pH conditions, they have been able to extract various combinations of phenolics from plants [13].

Plants contain various amounts of phenolic compounds which have simple and complicated structures. As there is a possibility of interaction between these compounds and other compounds in plants like carbohydrates and proteins, therefore it will be difficult to find an appropriate method for the extractions of all phenolic compounds [14]. In the liquid-liquid extraction (LLE) methods, various phenolic compounds are extracted and then an extra step for their purification is required.

There are three methods for extractions of phenolic compounds using LLE method which include Soxhlet extraction, maceration and hydro distillation methods. The effective parameters in these extraction methods are the type and polarity of solvents and their ratio, time and temperature of extraction and moreover chemical composition and physical characteristics of the samples [15].

Kerrouri et al. investigated the total content of phenolic compounds, flavonoids and condensed tannins with different solvents with different polarities using Soxhlet extractor and maceration methods [1]. Based on the results obtained in this paper, the various groups of phenolic compounds were extracted by solvents with different polarity so that hexane, ethyl acetate and methanol were used as best solvents for polyphenols, flavonoids and tannins respectively. Therefore they evaluated their antioxidant activities by their reaction with 1,1-diphenyl-2-picryl-hydrazyl radical. The results demonstrated that the antioxidants contained in the *Anethum graveolens* extracts were able to giving hydrogen to a free radical to eliminate potential damage. In another work, the separation of phenolic compounds from oil mixtures by three imidazolium-based dicationic ionic liquids was investigated by Wu et al. [16]. 1,4-bis[N-(N′-methylimidazolium)]butane dibromide demonstrated the maximum phenol removal yield of 96.6% within 5 min. The reusability of dicationic ionic liquids structures were investigated and indicated they were stable after four cycles without a decrease in phenol removal efficiency.

In the liquid-liquid extraction, usually 1–30 g of phenolic compounds within 6–24 h can be extracted. The advantages of this method are that, they have simple extraction procedures and various phenolic compounds can be extracted by organic solvents with different polarities but the disadvantages of this procedure are that they suffer from high solvent consumption, long extraction times and low extraction yields, exposure risk to organic vapors and degradation of target compounds during the extraction method [15]. The problems with these extraction methods are caused to creation and development of alternative techniques that in the following of this chapter will be explained.

3. Ultrasound-assisted extraction (UAE)

Ultrasound waves occur at frequencies between 20 kHz and 10 MHz which pass through solid, liquid and gas and also human is not able to hear it. In this extraction method, cavitation bubbles are created near the sample tissue then they break the cell wall and therefore cell content is released [17].

Ultrasound waves are applied and used by two probe and bath systems for extraction of phenolic compounds from plants. Beside of the inherent parameters of ultrasonic devices (such as amplitude, frequency and wavelength), their power and intensity also have a lot of effects on extraction process that they need to be optimized. Design and shape of reactor and also shape of probe can effect on extraction process [18].

Comparing to traditional methods, ultrasound-assisted extraction (UAE) method has been considered according to its simplicity, easy handling, low cost, high efficiency, lower organic solvent consumption and reduced extraction time. It can be used as a simple and reliable procedure in extensive range of organic solvents for various phenolic compounds in large-scale level and industry [19].

When the extraction process is underway on large-scale, temperature, time and type of solvent can be effective not only on the extraction efficiency but also on extraction compositions and they must pay a lot of attention. Therefore precise study of these parameters in order to obtain the best extraction efficiency is so important at industry. Although it is worth to mention that, high efficiency is not just the only main issue of the extraction process but the less renewable energy sources consumption is also very important in large-scale level and industry [15].

Sun et al. reported the effects of acoustic energy density (6.8–47.4 W/L) and temperature (20–50°C) on the extraction yields of total phenolics by ultrasound-assisted extraction from grape marc [20]. They used 50% aqueous ethanol as the solvent. They mentioned that, the initial extraction rate and final extraction yield with the increase of acoustic energy density and temperature increase higher due to higher diffusion coefficients. Moreover, the comparison between ultrasound technology and two conventional extraction methods showed that ultrasound is a competitive and effective extraction technology for extracting phenolic compounds from grape marc. The extraction yield of selected phenolic acids and flavonoids from *Equisetum arvense* L. herb carried out by Oniszczuk et al. [21]. Different extraction methods like oxhlet extraction, ultrasound-assisted extraction (USAE), and accelerated solvent extraction (ASE) were used in this work. They have mentioned that, ultrasound assisted extraction at 60°C in three cycles for 30 min, with 80% aqueous solution of methanol was a more effective and accurate method than other methods for isolation of selected phenolic acids, and flavonoids from *E. arvense* L. herb. In another research plan by Palma et al., they investigated various parameters such as temperature, output amplitude, duty cycle, the quantity of sample and the total extraction time for extraction phenolic compounds from grapes and compared optimum conditions with traditional extraction techniques [22]. The result showed that ultrasound assisted extraction was able to extract phenolic compounds in higher yield and much shorter extraction time, 6 min instead of 60 min. Also the extraction of phenolic compounds from Syrah grape skin was done by the Tonon et al. [23]. They optimized ultrasound power, citric acid concentration and solid to liquid ratio for this extraction method. Under optimum conditions, 59% of the quantified phenolic compounds with only 3 min of processing were extracted which compared to conventional extraction, ultrasound was considered a suitable method based on facilitating release of phenolic compounds from matrix. In the same way, Zardo et al. have reported extraction of phenolic compounds from sunflower seed cake [24]. The temperature and ethanol concentration showed the highest effect on the total phenolic compounds extraction from sunflower cake. The results showed that high amounts of phenolics compounds were obtained in the first minute of extraction and longer time did not effect on amount of extraction. Also in the other research plan Row et al. performed extraction of phenolic compounds from Laminaria japonica Aresch with three kinds of 1-alkyl-3-methylimidazolium with different cations and anions [25]. The results showed that, the characteristics of both anions and cations have remarkable effects on the extraction efficiency. Comparing the results of extraction under optimal conditions with conventional solvent showed highest extraction efficiency within the shortest extraction time. Sheng et al. evaluated antioxidant activities of phenolic compounds extracted from *Terminalia chebula* Retz. fruits by different extraction methods like UAE and LLE [26]. The results showed that, the antioxidant activities of phenolic compounds extracted by UAE under optimized condition were stronger

Sample	Phenolic compound	Solvent	Time (min)	Temperature (°C)	mg compound/g dry sample	Ref.
Grape marc	—	50% aqueous ethanol	80	50	1.57 mg/g	[20]
Equisetum arvense L. herb	Chlorogenic acid (CGA), caffeic acid (CA), ferulic acid (FA), isoquercitrin (IQ), 5-glucoside luteolin (5GL)	80% aqueous methanol	30	60	0.5631 mg_{CGA}/g, 0.4739 mg_{CA}/g, 0.2120 mg_{FA}/g, 0.2629 mg_{IQ}/g, 0.2485 mg_{5GL}/g	[21]
Grapes	Phenolic compounds (PC), tannins (TA), anthocyanins (AN)	50% aqueous ethanol	6	10	13 mg_{PC}/g, 7.2 mg_{TA}/g, 1.5 mg_{AN}/g	[22]
Syrah grape skin	Phenolic compounds	50% aqueous ethanol	4–10	20–80	11,732 mg/100 g	[23, 31]
Areca husk	Gallic acid	41% aqueous ethanol	38	53	15.37 mg/g	[32]
Sunflower seed	Gallic acid (GA), chlorogenic acid (CA)	43% aqueous ethanol	3	70	1867.3 mg_{GA}/100 g, 1645.8 mg_{CA}/100 g	[24]
Olive leaves	oleuropein (OLE), verbascoside (VER), luteolin-40-O-glucoside (L4OG)	80% aqueous ethanol	1	60	13.386 mg_{OLE}/g, 0.363 mg_{VER}/g, 0.527 mg_{L4OG}/g	[33]
Flax seeds	Lignan (LI), flavonol (FL), hydroxycinnamic acids (HA)	water with 0.2 N of sodium hydroxide	60	25	24.07 mg_{LI}/g, 6.84 mg_{FL}/g, 11.25 mg_{HA}/g	[34]
Brosimum alicastrum leaf	Total phenolic content (TPC), total monomeric anthocyanin (TMA)	80% aqueous methanol	20–10	28	45.18 mg $_{GAE}$/g, 15.16 mg_{CyE}/100 g	[35]
Grapefruit leaves	Gallic acid	10.80% aqueous ethanol	58.52	30.37	19.04 mg/g	[36]

Table 2.
Optimized condition for extraction of phenolic compounds from plants using UAE.

than LLE method. Recently, Wang et al. reported extraction of hemicellulose and phenolic compounds from bamboo bast fiber powder [18]. The nature of extracted phenolic compounds depends on the used solvent and they reported extraction of hemicellulose and phenolic compounds using ultrasound without adding harmful solvents. The results demonstrated that, the efficiency of extractions increased by 2.6-fold for ultrasound treated samples without solvent in comparison with extractions that use water. In fact, ultrasound can be as used as a green technology for the extraction of bamboo components without adding solvent.

Based on various literatures about ultrasound-assisted extraction method, phenolic compounds can be extracted within 10–60 min and compared to the LLE method, which is easy to be handled, inexpensive, safe and reproducible and can be simultaneously used for a wide range of samples [27, 28]. Further in UAE method, less solvent is required which is a great importance for environment and economical point of views and makes the industry more inclined to use this method in large-scale (**Table 2**). These properties represent that ultrasound-assisted extraction method is beneficial alternative instead of LLE method for extraction of phenolic compounds from different part of plants [29, 30].

It should be mentioned that in this method, an additional filtration step is required. Also as other extraction methods, degradation of compounds is possible at high frequencies.

4. Microwave-assisted extraction (MAE)

Microwave is electromagnetic radiations with frequencies in the range of 30–300 MHz. They generate heat due to the induction of molecular motions, which causes the cell wall to be ruptured and active substance in cell is released.

Dipolar solvent molecules such as water with higher dielectric constant than non-polar solvents can absorb high energy therefore they increase the speed and efficiency of extraction phenolic compounds [37].

There are two microwave-assisted extraction (MAE) systems for extraction of phenolic compounds which are closed vessel and open vessel. In the closed vessel system the extraction happens in the high pressure and temperature while in the open vessel system, the extraction of phenolic compound occurs in atmospheric pressure [38].

Until now, many literatures have been reported about applications of microwave technology for extraction of phenolic compounds from plants. Reducing time and cost, high efficiency and lower organic solvents consumption are advantages of this method than traditional methods. Also another important benefit of this extraction procedure is simultaneous extractions of several substances in a short time [15].

Proestos and Komaitis et al. carried out extraction of phenolic substances from aromatic plants with different solvents by microwave assisted extraction [39]. Comparing these results with conventional (reflux) extraction showed that, extraction by microwave assisted extraction could perform in shorter extraction time, less solvent usage and better extraction yield. Also Memon et al. investigated extraction, identification and antioxidative properties of flavonoids of leaves and flowers of *Cassia angustifolia* [40]. The extraction of flavonoids performed by five various methods including microwave extraction, Soxhlet extraction, sonication extraction, marinated extraction and reflux condensation extraction. Comparison of total flavonoids content with various extraction methods showed that, microwave extraction method is the best option and can extract 28.15–26.3 mg/g in the flowers and leaves, respectively. The advantages of this method were that it was very easy, robust and the extraction was carried out in the 9 min, which is the shortest time required in comparison with other methods. Therefore quantity and antioxidant activity for this extraction method was found to be the more efficient for extracting more numbers of flavonoids.

Guolin et al. carried out microwave-assisted extraction of pectin from lemon peels using ionic liquid as alternative solvent [41]. Under the optimal conditions and 9.6 min duration time, the extraction efficiency of pectin was 24.68%, which is much higher than the efficiency from the conventional heating reflux extraction under 2 h duration. According to this result, ionic liquid solutions could be considered as an effective solvent in microwave-assisted extraction of pectin from lemon peels. Also Liu et al. evaluated extraction and determination of taxifolin *Larix gmelinii* by ionic liquid-based microwave-assisted extraction method (ILMAE) [31]. They investigated different kinds of 1-alkyl-3-methylimidazolium ionic liquids with different kinds of cations and anions. According to the results, both anions and cations parts of ionic liquids could affect to the extraction yields and optimal solvent was 1-butyl-3-methylimidazolium bromide for the taxifolin extraction. Comparing this method with traditional method including water stirring extraction (WSE), water reflux extraction (WRE), and maceration showed that, the ILMAE method could be done in higher extraction yield, lower energy and time consumption. In the following, under optimum extraction conditions, taxifolin was stable and there was no degradation on it.

Antioxidant activity and inhibitory effect of phenolic compounds extracted by microwave assisted extraction method, are higher and more efficient than

Sample	Phenolic compound	Solvent	Time (min)	Temperature (°C)	mg compound/g dry sample	Ref.
Leaves and flowers of *Cassia angustifolia*	Flavonoids	70% aqueous ethanol	9	50	28.15 $mg_{flowers}$/g, 26.30 mg_{leaves}/g	[40]
Medicinal plants	Flavonoids (FV) Phenolic acids (PA)	50% aqueous ethanol	3–5	70	0.84 g_{FV}/100 g, 2.01 g_{PA}/100 g	[2]
Green coffee beans	Chlorogenic acids	Water	5	50	23.93 mg/g	[47]
Morus nigra leaves	Gallic acid	50% aqueous ethanol	28	120	19.7 mg/g	[48]
Olive Leaf	Phenolic compounds	80% aqueous methanol	6	80	—	[49]
Cherry laurel fruit	Chlorogenic acid (CA), vanillic acid (VA)	Methanol	12–16	65	35.21 mg_{CA}/g 1.19 mg_{VA}/g	[50]
Cherry laurel leaf	Chlorogenic acid (CA), luteolin 7-glucoside (L7G)	Methanol	12–30	65	30.44 mg_{CA}/g 22.78 mg_{L7G}/g	[50]
Larix gmelinii	Taxifolin	[C_4mim]Br	14	—	18.47 mg/g	[31]
Psidium guajava Linn leaves	Gallic acid (GA), ellagic acid (EA), quercetin (Q)	Ionic liquid	10	70	0.507 mg_{GA}/g 2.387 mg_{EA}/g 0.540 mg_{Q}/g	[51]
Smilax china tubers	*trans*-Resveratrol (*t*-R), quercetin (Q)	Ionic liquid	10	60	0.531 $mg_{t\text{-}R}$/g 0.189 mg_{Q}/g	[51]

Table 3.
Optimized condition for extraction of phenolic compounds from plants using MAE.

LLE method because a shorter time is required in microwave assisted extraction. According to this fact, Yuan et al. reported the extraction of phenolic compounds at 110°C for 15 min from four economic brown macroalgae species using MAE. These extracted compounds had a higher antioxidant activity and total phenolic content for all algae species in comparison with those obtained by conventional extraction at room temperature for 4 h [42]. In another report from Alara et al., phenolic compounds were extracted from Vernoniacinerea leaves by microwave assisted extraction and they had antioxidant and anti-diabetic activity more than LLE method [43].

Overall, effective parameters of this extraction procedure are solvent, pressure, temperature, nature of matrix and power of device. In microwave-assisted extraction method compared to the LLE, it is easy to handle and less solvent is required [44–46]. Further, this method can reduce the extraction time and increase the extraction yield. The dark point of this method is that, solvent must absorb microwave power and also there is a risk of explosion in this method (**Table 3**).

5. Supercritical fluid extraction (SFE)

In this method, the solvent is at a temperature and pressure above its critical point and there is no surface tension in it. Therefore it simultaneously has the properties of liquid and gas which can be much efficient for extraction of phenolic compounds from plants. Low viscosities and high diffusivities of supercritical fluids enable them to extract the various phenolic compounds in less time with higher efficiency. So it's a good alternative instead of using liquid-liquid extraction methods [52]. A distinguished property of supercritical fluid extraction is that, the density of supercritical fluid can be easily changed based on various temperatures and pressures. Also in constant temperature, solubility in a supercritical fluid is directly related to its density so that by increasing pressure, its solubility tends to increase. Hence, these properties can be used to separate phenolic compounds with various selectivity [53].

Carbon dioxide is known as a stable, nontoxic, environmentally safe, cheap, and selective extraction solvent for supercritical fluid extraction. Based on these outstanding properties, many literatures have been published that carbon dioxide has been used as solvent for extraction of various phenolic compounds [54, 55]. Murga et al. reported selective extraction of some phenolic compounds from grape seeds using carbon dioxide and alcohol under supercritical conditions [52]. However, carbon dioxide is a non-polar solvent and to achieve higher efficiency and selectivity in phenolic compounds extraction, they adjusted the polarity of carbon dioxide by adding methanol and ethanol on it. Various phenolic compounds under optimized conditions were extracted which are mentioned in **Table 4**. Vatai et al. investigated extraction of phenolic compounds from elder berry by supercritical carbon dioxide [54]. They optimized pressure, temperature and carbon dioxide/ethanol mixture ratio for this extraction method and compared the results with liquid-liquid extraction method. The results demonstrated that, supercritical extraction with carbon dioxide can improve the extraction of total phenols from elder berry and hence, it is a good alternative procedure to replace instead of organic solvents.

The antioxidant capacities of phenolic compounds extracted by supercritical fluid extraction are much higher than LLE method. Recently, Pereira et al. assessed antioxidant activity of bioactive compounds which were extracted from myrtle leaves and berries by supercritical fluid extraction method [56]. They used three different methods for investigation of antioxidant activity of these compounds and the results demonstrated that, the extracted phenolic compounds by supercritical fluid extraction methods have a higher antioxidant activity than LLE method. Also, Tan et al. reported that, antioxidant capacity of virgin avocado oil extracted

Sample	Phenolic compound	Pressure (bar)	Time (min)	Temperature (°C)	mg compound/g dry sample	Ref.
Defatted milled grape seeds	gallic acid (GA), protocatechuic acid (PCA), monogalloyl glucose (MG), protocatechualdehyde (PRA), (+)catechin(C), (−) epicatechin (E)	300	180	40	0.034 mg_{GA}/g 0.015 mg_{PCA}/g 0.002 mg_{MG}/g 0.003 mg_{PRA}/g 0.058 mg_{C}/g 0.038 mg_{E}/g	[52]
Elder Berry	Total phenols, quercetin (Q), *t*-resveratrol (*t*-R)	150–300	—	40	74.6 mg_{GA}/g 152 $mg_{Q}/100$ g 21.0 $mg_{tR}/100$ g	[54]
Vitis labrusca B	Total phenols (TPC), anthocyanins (A)	157–161	—	45–46	2.156 $mg_{TPC}/100$ mL 1.176 mg_{A}/mL	[59]
Olive leaves	Phenolic compounds	334	140	100	4.2 mg/g	[53]
Agricultural by-product (wheat straw)	Lignin derived bioactive compounds e.g., tricin and catechins	140	—	35	—	[60]
Eugenia uniflora	Phenolic compounds	400	360	60	240.5 mg_{GAE}/g	[61]
Guaraná seeds	Phenolic compounds (pyrogallol)	100	40	40	105.76 mg/g	[55]
Sasa palmate	Gluconic acid, β-siosterol, α-amyrin acetate	200	—	95	7.31 mg/g	[62]

Table 4.
Optimized conditions for extraction of phenolic compounds from plants using SFE.

using supercritical fluid extraction was higher than UAE and LLE methods [57]. Moreover, avocado oil extracted using supercritical fluid extraction had two to four times bigger levels of α- and γ-tocopherols than LLE and UAE.

In conclusion, the supercritical fluid extraction method is rapid, inexpensive to run and selective, so that even a small available amount of phenolic compounds in plants can be extracted using this method. Also, this method requires a small amount of solvent or no solvent and thermally labile compounds will be stable during this extraction method [58].

6. The purification and identification of phenolic compounds

A very large number of literatures have been published about separation, purification and identification of phenolic compounds in plants. According to the importance of this matter, many improvements have been achieved in this field. These protocols and procedures can be used for identifying and isolating the compounds with a high precision. Since various parts of plants have phenolic compounds with different structures and chemical properties, so that various protocols and spectroscopic techniques such as UV-visible spectroscopy, near-infrared reflectance spectroscopy, nuclear magnetic resonance and high performance liquid chromatography (HPLC) have to be used for their purifications and identifications.

6.1 UV-visible spectroscopy

This spectroscopic method can be used for measuring absorption spectroscopy in the ultraviolet-visible spectral region which relates to electronic transitions in molecules. Based on aromatic structure of phenolic compounds, they are powerful chromophores in the UV rang so that flavones, phenolic acids and total anthocyanins have absorptions spectrums in the region 320, 360, and 520 nm respectively [63]. The advantages of this spectroscopic method are less time consuming, inexpensive and reproducible. The limitations of identification of this method get raised when multiple combinations have partially overlapping peaks. In order to solve this problem, the method must be used with other spectroscopic methods such as mass spectroscopy or HPLC, simultaneously [64].

6.2 Infrared spectroscopy

This spectroscopic method uses a wide range of wavelength from 780 nm to 1 mm. According to its wavelength, it can be divided to three regions; the near-, mid- and far-infrared. These electromagnetic wavelengths cause vibrational changes in the molecules. Therefore, it is an efficient method for identification of molecules and especially their functional groups, so that various functional groups such as single bond, double bond, triple bond, carboxyl, hydroxyl and amino groups have diverse vibrational frequencies. In fact, with this spectroscopy method, we cannot draw the structure of a chemical compounds and we can just understand the nature of the functional groups in the structures [63]. Abbas et al. investigated the mid-infrared spectra of 36 standard phenolic compounds. The survey of the MIR spectra at the 1755–1400 cm^{-1} and 1000–870 cm^{-1} regions for these compounds showed that, MIR method was able to distinguish between flavonoids and phenolic acids families. They finally mentioned that, in order to have a better distinguishment between various families of phenolic compounds it is necessary to investigate a larger number of samples [65].

6.3 Nuclear magnetic resonance

It is a technique based on the measurement of electromagnetic radiation in the radio waves area. Unlike the infrared and UV-visible spectroscopic methods which electrons are involved in absorption process, this technique is related to the magnetic properties of atomic nuclei. Commonly ^{1}H NMR and ^{13}C NMR are used to study the chemical structure of materials but it can be also applicable for any kind of nuclei possessing spin. Using this spectroscopy method, in addition to identifying chemical compounds, and accurate information on the structure, dynamics, reaction state and chemical environment of molecules can be provided too [66]. Many articles have been published in this regard demonstrating that ^{1}H NMR spectroscopy can be an effective and a useful method for identifying and analyzing of phenolic compounds in plants [63, 67]. Furthermore, this method is fast, quantitative and non-destructive.

6.4 High performance liquid chromatography

This technique is one of the most efficient and impressive chromatographic method that has widely been used for separation, identification and quantification of phenolic compounds. In this method, using the mobile phase at high pressure, sample mixture will be separated from each other on stationary phase. Generally, for the separation and quantification of phenolic compounds, the HPLC is preferred than gas chromatography (GC). Various factors can effect to the efficiency of HPLC which include sample purification, mobile phase, column types, and detectors [33]. Various literatures have been published regarding purifying and quantifying phenolic compounds using HPLC technique [49, 68]. Normally, various phenolic compounds can be separated using normal phase C18 or reversed phase (RP-C18) column in the presence of different solvents with different polarities. Also it should be mentioned that pH mobile phase should be stabilized in the range of 2–4 to avoid the ionization of phenolic compounds. Therefore, aqueous acidified mobile phase has to be used during purification and quantification of phenolic compounds [23].

In the HPLC technique, the selection of detector plays a significant role in the identification process and depends on the properties of the phenolic compounds. High performance liquid chromatography-mass spectrometry, (HPLC-MS), is highly selective and has a low limit of detection and shorter analysis time [40]. Today, this method is being extensively used because of it is enormous benefits. La Torre-Carbot et al. investigated characterization and quantification of 20 phenolic compounds in olive oils through a combination of the HPLC-DAD and HPLC-MS/MS methods [69]. This method was fast, precise and sensitive. Also this method required a low solvent and sample consumption. HPLC-UV can be used for phenolic compounds that have UV spectrum and thus, selection of wavelength is the most important point for this method so that the solvent and extra available compounds should not have absorption in this wavelength [51]. HPLC with fluorescence detection. (HPLC-FLD), is an efficient method for identification and quantification of phenolic compounds that have fluorescence spectrum. The fluorescence detector sensitivity is 10–1000 times higher than UV detector [70]. HPLC with electrochemical detection is based on redox reaction of phenolic compounds. This method is fast, low cost, precise and low limits of detection of phenolic compounds [71]. Cantalapiedra et al. have used HPLC for separating vanillin, eugenol, thymol and carvacrol using amperometric and coulometric detectors [72]. They reported that, the coulometric detection has a low limit of detection in the range between 0.81 and 3.1 μg/L and is very competitive and sufficient for quality control of phenolic compounds in comparison with other methods, such as GC-MS which are expensive and complicated.

7. Conclusions

In summary, phenolic compounds extracted from plants, have different applications as antioxidant, antimicrobial, anti-inflammatory, antitumor, antiviral, analgesic and antipyretic. Therefore, they have this ability to improve human health. In this chapter, various extraction methods of phenolic compounds from plants were presented and their advantages and disadvantages were explained. According to the structure and extraction source of these compounds, different extraction methods can be recommended. Suggested methods need to be simple and rapid with a high. Also an important point for selecting an extraction method is that, it should be environmentally friendly. In addition, in the last part of this chapter various methods for purification and identification of phenolic compounds were presented with their advantages and disadvantages.

Acknowledgements

I would like to thank Alexander von Humboldt Foundation to financial support to achieve this work.

Author details

Shadab Gharaati Jahromi
Institute of Inorganic Chemistry and Analytical Chemistry, Johannes Gutenberg University Mainz, Mainz, Germany

*Address all correspondence to: shadabgharaati@gmail.com

References

[1] Cheynier V. Phenolic compounds: From plants to foods. Phytochemistry Reviews. 2012;**11**(2-3):153-177

[2] Salinas-Moreno Y, García-Salinas C, Ramírez-Díaz JL, La Alemán-de Torre I. Phenolic compounds in maize grains and its nixtamalized products. In: Soto-Hernández M, Palma-Tenango M, Garcia-Mateos MDR, editors. Importance and Applications. Phenolic Compounds—Natural Sources. IntechOpen; 2017

[3] Lin D, Xiao M, Zhao J, Li Z, Xing B, Li X, et al. An overview of plant phenolic compounds and their importance in human nutrition and management of type 2 diabetes. Molecules (Basel, Switzerland). January 2016;**21**(10)

[4] Gaytán VG, Mendoza MNR, Hernández MS, Trejo-Téllez LI, Santos MEP, Ponce GV. Volatile components in the flower, pedicellate ovary and aqueous residue of cymbidium sp. (ORCHIDACEAE). Journal of Analytical Sciences,Methods and Instrumentation. 2013;**03**(04):212-218

[5] Rios TS, Torres TS, Cerrilla MEO, Hernández MS, Cruz AD, Bautista JH, et al. Changes in composition, antioxidant content, and antioxidant capacity of coffee pulp during the ensiling process. Revista Brasileira de Zootecnia. 2014;**43**(9):492-498

[6] Gharaati S, Kargar H, Falahati AM. Tetrahydropyranylation of alcohols and phenols catalyzed by a new multi-wall carbon nanotubes-bound tin(IV) porphyrin. Journal of the Iranian Chemical Society. 2017;**14**(6):169-1178

[7] Pereira D, Valentão P, Pereira J, Andrade P. Phenolics: From chemistry to biology. Molecules. 2009;**14**(6):2202-2211

[8] Moghadam M, Tangestaninejad S, Mirkhani V, Mohammadpoor-Baltork I, Gharaati S. Polystyrene-bound electron-deficient tin(IV) porphyrin: A new, highly efficient, robust and reusable catalyst for acetylation of alcohols and phenols with acetic anhydride. Comptes Rendus Chimie. 2011;**14**(12):1080-1087

[9] Gharaati S, Moghadam M, Tangestaninejad S, Mirkhani V, Mohammadpoor-Baltork I, Kosari F. Highly efficient and selective methoxymethylation of alcohols and phenols catalyzed by high-valent tin(IV) porphyrin. Inorganica Chimica Acta. 2010;**363**(9):1995-2000

[10] Cheynier V, Comte G, Davies KM, Lattanzio V, Martens S. Plant phenolics: Recent advances on their biosynthesis, genetics, and ecophysiology. Plant Physiology and Biochemistry: PPB. 2013;**72**:1-20

[11] Gharaati S, Moghadam M, Tangestaninejad S, Mirkhani V, Mohammadpoor-Baltork I. Highly efficient and selective trimethylsilylation of alcohols and phenols with hexamethyldisilazane catalyzed by polystyrene-bound tin(IV) porphyrin. Polyhedron. 2012;**35**(1):87-95

[12] Dai J, Mumper RJ. Plant phenolics: Extraction, analysis and their antioxidant and anticancer properties. Molecules (Basel, Switzerland). 2010;**15**(10):7313-7352

[13] Jiao T, Li C, Zhuang X, Cao S, Chen H, Zhang S. The new liquid-liquid extraction method for separation of phenolic compounds from coal tar. Chemical Engineering Journal. 2015;**266**:148-155

[14] Yu J, Ahmedna M, Goktepe I. Effects of processing methods and extraction solvents on concentration and antioxidant activity of peanut skin phenolics. Food Chemistry. 2005;**90**(1-2):199-206

[15] Garcia-Salas P, Morales-Soto A, Segura-Carretero A, Fernández-Gutiérrez A. Phenolic-compound-extraction systems for fruit and vegetable samples. Molecules. 2010;**15**(12):8813-8826

[16] Ji Y, Hou Y, Ren S, Yao C, Wu W. Highly efficient separation of phenolic compounds from oil mixtures by imidazolium-based dicationic ionic liquids via forming deep eutectic solvents. Energy & Fuels. 2017;**31**(9):10274-10282

[17] Altemimi A, Watson DG, Choudhary R, Dasari MR, Lightfoot DA. Ultrasound assisted extraction of phenolic compounds from peaches and pumpkins. PLoS ONE. 2016;**11**(2):e0148758

[18] Wang C, Tallian C, Su J, Vielnascher R, Silva C, Cavaco-Paulo A, et al. Ultrasound-assisted extraction of hemicellulose and phenolic compounds from bamboo bast fiber powder. PLoS ONE. 2018;**13**(6):e0197537

[19] Olmo-García L, Bajoub A, Benlamaalam S, Hurtado-Fernández E, Bagur-González MG, Chigr M, et al. Establishing the phenolic composition of *Olea europaea* L. leaves from cultivars grown in morocco as a crucial step towards their subsequent exploitation. Molecules (Basel, Switzerland). 2018;**23**(10)

[20] Tao Y, Zhang Z, Sun DW. Kinetic modeling of ultrasound-assisted extraction of phenolic compounds from grape marc: Influence of acoustic energy density and temperature. Ultrasonics Sonochemistry. 2014;**21**(4):1461-1469

[21] Oniszczuk A, Podgórski R, Oniszczuk T, Żukiewicz-Sobczak W, Nowak R, Waksmundzka-Hajnos M. Extraction methods for the determination of phenolic compounds from *Equisetum arvense* L. herb. Industrial Crops and Products. 2014;**61**:377-381

[22] Carrera C, Ruiz-Rodríguez A, Palma M, Barroso CG. Ultrasound assisted extraction of phenolic compounds from grapes. Analytica Chimica Acta. 2012;**732**:100-104

[23] Mazza KEL, Santiago MCPA, do Nascimento LSM, Godoy RLO, Souza EF, Brígida AIS, et al. Syrah grape skin valorisation using ultrasound-assisted extraction: Phenolic compounds recovery, antioxidant capacity and phenolic profile. International Journal of Food Science and Technology. 2018;**34**:206

[24] Zardo I, Espíndola Sobczyk A, de Marczak LDF, Sarkis J. Optimization of ultrasound assisted extraction of phenolic compounds from sunflower seed cake using response surface methodology. Waste and Biomass Valorization. 2017;**50**:1713

[25] Han D, Zhu T, Row KH. Ultrasonic extraction of phenolic compounds from *Laminaria japonica* aresch using ionic liquid as extraction solvent. Bulletin of the Korean Chemical Society. 2011;**32**(7):2212-2216

[26] Sheng Z, Zhao J, Muhammad I, Zhang Y. Optimization of total phenolic content from *Terminalia chebula* Retz. Fruits using response surface methodology and evaluation of their antioxidant activities. PLoS ONE. 2018;**13**(8):e0202368

[27] Turrini F, Boggia R, Leardi R, Borriello M, Zunin P. Optimization of the ultrasonic-assisted extraction of phenolic compounds from *Oryza Sativa* L. 'Violet Nori' and determination of the antioxidant properties of its caryopses and leaves. Molecules. 2018;**23**(4)

[28] Aware CB, Patil RR, Vyavahare GD, Gurme ST, Jadhav JP. Ultrasound-assisted aqueous extraction of phenolic, flavonoid compounds and antioxidant activity of *Mucuna macrocarpa* beans: Response surface methodology

optimization. Journal of the American College of Nutrition. 2018:1-9

[29] Abdelkebir R, Alcántara C, Falcó I, Sánchez G, Garcia-Perez JV, Neffati M, et al. Effect of ultrasound technology combined with binary mixtures of ethanol and water on antibacterial and antiviral activities of Erodium glaucophyllum extracts. Innovative Food Science & Emerging Technologies. 2019;**52**:189-196

[30] Ismail BB, Guo M, Pu Y, Wang W, Ye X, Liu D. Valorisation of baobab (*Adansonia digitata*) seeds by ultrasound assisted extraction of polyphenolics. Optimisation and comparison with conventional methods. Ultrasonics Sonochemistry. 2019;**52**:257-267

[31] Liu Z, Jia J, Chen F, Yang F, Zu Y, Yang L. Development of an ionic liquid-based microwave-assisted method for the extraction and determination of taxifolin in different parts of *Larix gmelinii*. Molecules (Basel, Switzerland). 2014;**19**(12):19471-19490

[32] Chen W, Huang Y, Qi J, Tang M, Zheng Y, Zhao S, et al. Optimization of ultrasound-assisted extraction of phenolic compounds from areca husk. Journal of Food Processing & Preservation. 2014;**38**(1):90-96

[33] Giacometti J, Žauhar G, Žuvić M. Optimization of ultrasonic-assisted extraction of major phenolic compounds from olive leaves (*Olea europaea* L.) using response surface methodology. Foods (Basel, Switzerland). 2018;**7**(9)

[34] Corbin C, Fidel T, Leclerc EA, Barakzoy E, Sagot N, Falguiéres A, et al. Development and validation of an efficient ultrasound assisted extraction of phenolic compounds from flax (*Linum usitatissimum* L.) seeds. Ultrasonics Sonochemistry. 2015;**26**:176-185

[35] Gullian Klanian M, Terrats Preciat M. Optimization of the ultrasound-assisted extraction of phenolic compounds from *Brosimum alicastrum* leaves and the evaluation of their radical-scavenging activity. Molecules (Basel, Switzerland). 2017;**22**(8)

[36] Ciğeroğlu Z, Aras Ö, Pinto CA, Bayramoglu M, Kırbaşlar Şİ, Lorenzo JM, et al. Optimization of ultrasound-assisted extraction of phenolic compounds from grapefruit (*Citrus paradisi* Macf.) leaves via D-optimal design and artificial neural network design with categorical and quantitative variables. Journal of the Science of Food and Agriculture. 2018;**98**(12):4584-4596

[37] Sookjitsumran W, Devahastin S, Mujumdar AS, Chiewchan N. Comparative evaluation of microwave-assisted extraction and preheated solvent extraction of bioactive compounds from a plant material: A case study with cabbages. International Journal of Food Science and Technology. 2016;**51**(11):2440-2449

[38] Vlaisavljević S, Kaurinović B, Popović M, Vasiljević S. Profile of phenolic compounds in *Trifolium pratense* L. extracts at different growth stages and their biological activities. International Journal of Food Properties. 2017;**20**(12):3090-3101

[39] Proestos C, Komaitis M. Ultrasonically assisted extraction of phenolic compounds from aromatic plants: Comparison with conventional extraction technics. Journal of Food Quality. 2006;**29**(5):567-582

[40] Laghari AQ, Memon S, Nelofar A, Laghari AH. Extraction, identification and antioxidative properties of the flavonoid-rich fractions from leaves and flowers of *Cassia angustifolia*. AJAC. 2011;**02**(08):871-878

[41] Guolin H, Jeffrey S, Kai Z, Xiaolan H. Application of ionic liquids in the

microwave-assisted extraction of pectin from lemon peels. Journal of Analytical Methods in Chemistry. 2012;**2012**:302059

[42] Yuan Y, Zhang J, Fan J, Clark J, Shen P, Li Y, et al. Microwave assisted extraction of phenolic compounds from four economic brown macroalgae species and evaluation of their antioxidant activities and inhibitory effects on α-amylase, α-glucosidase, pancreatic lipase and tyrosinase. Food Research International (Ottawa, Ont.). 2018;**113**:288-297

[43] Alara OR, Abdurahman NH, Ukaegbu CI, Azhari NH. Vernonia cinerea leaves as the source of phenolic compounds, antioxidants, and anti-diabetic activity using microwave-assisted extraction technique. Industrial Crops and Products. 2018;**122**:533-544

[44] Özbek HN, Halahlih F, Göğüş F, Koçak Yanık D, Azaizeh H. Pistachio (*Pistacia vera* L.) Hull as a potential source of phenolic compounds: Evaluation of ethanol–water binary solvent extraction on antioxidant activity and phenolic content of pistachio hull extracts. Waste and Biomass Valorization. 2018;**112**(2):394

[45] Hernández YR, García Serrano LA, Maruri DT, Jiménez Aparicio AR, Camacho Díaz BH, Arenas Ocampo ML. Optimization of the microwave-assisted ethanosolv extraction of lignocellulosic compounds from the bagasse of agave angustifolia haw using the response methodology. Journal of Agricultural and Food Chemistry. 2018;**66**(13):3533-3540

[46] Zhao CN, Zhang JJ, Li Y, Meng X, Li HB. Microwave-assisted extraction of phenolic compounds from *Melastoma sanguineum* fruit: Optimization and identification. Molecules. 2018;**23**(10):2498

[47] Upadhyay R, Ramalakshmi K, Jagan Mohan Rao L. Microwave-assisted extraction of chlorogenic acids from green coffee beans. Food Chemistry. 2012;**130**(1):184-188

[48] Radojković M, Moreira MM, Soares C, Fátima Barroso M, Cvetanović A, Švarc-Gajić J, et al. Microwave-assisted extraction of phenolic compounds from *Morus nigra* leaves: Optimization and characterization of the antioxidant activity and phenolic composition. Journal of Chemical Technology and Biotechnology. 2018;**93**(6):684-1693

[49] Taamalli A, Arráez-Román D, Ibañez E, Zarrouk M, Segura-Carretero A, Fernández-Gutiérrez A. Optimization of microwave-assisted extraction for the characterization of olive leaf phenolic compounds by using HPLC-ESI-TOF-MS/IT-MS 2. Journal of Agricultural and Food Chemistry. 2012;**60**(3):791-798

[50] Karabegović IT, Stojičević SS, Veličković DT, Todorović ZB, Nikolić NČ, Lazić ML. The effect of different extraction techniques on the composition and antioxidant activity of cherry laurel (*Prunus laurocerasus*) leaf and fruit extracts. Industrial Crops and Products. 2014;**54**:142-148

[51] Du FY, Xiao XH, Luo XJ, Li GK. Application of ionic liquids in the microwave-assisted extraction of polyphenolic compounds from medicinal plants. Talanta. 2009;**78**(3):1177-1184

[52] Murga R, Ruiz R, Beltrán S, Cabezas JL. Extraction of natural complex phenols and tannins from grape seeds by using supercritical mixtures of carbon dioxide and alcohol. Journal of Agricultural and Food Chemistry. 2000;**48**(8):3408-3412

[53] Le Floch F, Tena MT, Ríos A, Valcárcel M. Supercritical fluid extraction of phenol compounds from olive leaves. Talanta. 1998;**46**(5):1123-1130

[54] Vatai T, Škerget M, Knez Ž. Extraction of phenolic compounds from elder berry and different grape marc varieties using organic solvents and/or supercritical carbon dioxide. Journal of Food Engineering. 2009;**90**(2):246-251

[55] Marques LL, Panizzon GP, Aguiar BA, Simionato AS, Cardozo-Filho L, Andrade G, et al. Guaraná (*Paullinia cupana*) seeds: Selective supercritical extraction of phenolic compounds. Food Chemistry. 2016;**212**:703-711

[56] Pereira P, Cebola MJ, Oliveira MC, Bernardo-Gil MG. Supercritical fluid extraction vs conventional extraction of myrtle leaves and berries: Comparison of antioxidant activity and identification of bioactive compounds. The Journal of Supercritical Fluids. 2016;**113**:1-9

[57] Tan CX, Chong GH, Hamzah H, Ghazali HM. Characterization of virgin avocado oil obtained via advanced green techniques. European Journal of Lipid Science and Technology. 2018;**120**(10):1800170

[58] Khaw KY, Parat MO, Shaw PN, Falconer JR. Solvent supercritical fluid technologies to extract bioactive compounds from natural sources: A review. Molecules. 2017;**22**(7)

[59] Ghafoor K, Park J, Choi YH. Optimization of supercritical fluid extraction of bioactive compounds from grape (*Vitis labrusca* B.) peel by using response surface methodology. Innovative Food Science & Emerging Technologies. 2010;**11**(3):485-490

[60] da Costa Lopes AM, Brenner M, Falé P, Roseiro LB, Bogel-Łukasik R. Extraction and purification of phenolic compounds from lignocellulosic biomass assisted by ionic liquid, polymeric resins, and supercritical CO_2. ACS Sustainable Chemistry & Engineering. 2016;**4**(6):3357-3367

[61] Garmus TT, Paviani LC, Queiroga CL, Magalhães PM, Cabral FA. Extraction of phenolic compounds from Pitanga (*Eugenia uniflora* L.) leaves by sequential extraction in fixed bed extractor using supercritical CO_2, ethanol and water as solvents. The Journal of Supercritical Fluids. 2014;**86**:4-14

[62] Zulkafli ZD, Wang H, Miyashita F, Utsumi N, Tamura K. Cosolvent-modified supercritical carbon dioxide extraction of phenolic compounds from bamboo leaves (*Sasa palmata*). The Journal of Supercritical Fluids. 2014;**94**:123-129

[63] Bucar F, Wube A, Schmid M. Natural product isolation—How to get from biological material to pure compounds. Natural Product Reports. 2013;**30**(4):525-545

[64] Harnly JM, Bhagwat S, Lin LZ. Profiling methods for the determination of phenolic compounds in foods and dietary supplements. Analytical and Bioanalytical Chemistry. 2007;**389**(1):47-61

[65] Abbas O, Compère G, Larondelle Y, Pompeu D, Rogez H, Baeten V. Phenolic compound explorer: A mid-infrared spectroscopy database. Vibrational Spectroscopy. 2017;**92**:111-118

[66] Christophoridou S, Dais P. Novel approach to the detection and quantification of phenolic compounds in olive oil based on 31 P nuclear magnetic resonance spectroscopy. Journal of Agricultural and Food Chemistry. 2006;**54**(3):656-664

[67] Christophoridou S, Dais P. Detection and quantification of phenolic compounds in olive oil by high resolution 1H nuclear magnetic resonance spectroscopy. Analytica Chimica Acta. 2009;**633**(2):283-292

[68] Pérez-Serradilla JA, Priego-Capote F, Luque de Castro MD. Simultaneous

ultrasound-assisted emulsification-extraction of polar and nonpolar compounds from solid plant samples. Analytical Chemistry. 2007;**79**(17):6767-6774

[69] La Torre-Carbot K, de Jauregui O, Gimeno E, Castellote AI, Lamuela-Raventós RM, López-Sabater MC. Characterization and quantification of phenolic compounds in olive oils by solid-phase extraction, HPLC-DAD, and HPLC-MS/MS. Journal of Agricultural and Food Chemistry. 2005;**53**(11):4331-4340

[70] Tomaz I, Maslov L. Simultaneous determination of phenolic compounds in different matrices using phenyl-hexyl stationary phase. Food Analytical Methods. 2016;**9**(2):401-410

[71] Gazdik Z, Reznicek V, Adam V, Zitka O, Jurikova T, Krska B, et al. Use of liquid chromatography with electrochemical detection for the determination of antioxidants in less common fruits. Molecules (Basel, Switzerland). 2008;**13**(11):2823-2836

[72] Cantalapiedra A, Gismera MJ, Sevilla MT, Procopio JR. Sensitive and selective determination of phenolic compounds from aromatic plants using an electrochemical detection coupled with HPLC method. Phytochemical analysis: PCA. 2014;**25**(3):247-254

Chapter 2

Optimization of the Extraction of Polyphenols and Antioxidant Capacity from *Byrsonima crassifolia* (L.) Kunth Fruit by Response Surface Methodology

Mariana Séfora Bezerra Sousa,
Jovan Marques Lara Júnior and Diego de Souza Buarque

Abstract

The purpose of this research was to optimize the extraction conditions of polyphenols from murici (*Byrsonima crassifolia* (L.) Kunth) using the response surface methodology. Temperature (from 10 to 70°C), acetone concentration (from 10 to 100%), extraction time (from 0 to 160 min), and solid-liquid ratio (from 20 to 140 mg/mL) were investigated as independent variables in order to obtain the optimal conditions for extraction and to maximize the total phenolic content (TPC) and antioxidant activity (DPPH) of obtained extracts. Experimental results were fitted to the second-order polynomial model where multiple regression and analysis of variance were used to determine the fitness of the model and optimal condition for investigated responses. The solid-liquid did not interfere in the two responses. The results showed that for TPC extraction, the optimal conditions were obtained with an acetone concentration of 44%, a temperature of 29°C, and an extraction time of 51 min. For DPPH, the optimal conditions were the following: an acetone concentration of 45%, a temperature of 40°C, and an extraction time of 53 min. The use of such conditions allowed the maximum extraction of antioxidant murici at a lower cost of production, which may contribute to large-scale industrial applications and future pharmacological research.

Keywords: murici, total phenols, antioxidant activity, response surface methodology

1. Introduction

Byrsonima crassifolia (L.) Kunth, murici, is a species of fruit trees belonging to Malpighiaceae family, *Byrsonima* genus, which includes more than 150 species [1]. Generally, it occurs in the Amazon region and in the Brazilian Northeast [2].

Byrsonima crassifolia fruits are good sources of dietary fiber [3], vitamin C (84 mg 100 g^{-1}) [1], phenolic compounds, and carotenoids [4]. The main phenolics contained in murici include gallic acid and quercetin [5], and the main carotenoids, lutein and zeaxanthin [6]. In addition, there are reports of trace quantities of catechin, epicatechin, rutin, and kaempferol [7].

Some investigations have demonstrated the potential antioxidant effect of murici leaves, fruits, and seeds *in vitro* [5, 7, 8] and *in vivo* [9, 10]. The extracts can also be used as an alternative for the treatment of diabetes, reducing blood glucose, cholesterol, and triglycerides [11].

On the other hand, research suggests that the increased consumption of foods rich in antioxidants compounds, as polyphenols, is associated with a reduction of chronic disease risk [12, 13]. Polyphenols are an important group of phytochemicals originated from the secondary metabolism of plants. They are involved in plant growth and reproduction, pigmentation, and as agents that offer resistance to pathogens and stress conditions [14]. Moreover, the use of natural antioxidants, as polyphenols, has also been increased in the food and pharmaceutical industry. This increase is justified because the food industry has sought to replace synthetic antioxidants since that some of them present toxic activity [15].

However, there is no standard and universally accepted method for extracting antioxidant compounds. The methodological problems are due to great diversity of compounds, which give them different physicochemical properties, and their high susceptibility to oxidation [16]. This challenge is even greater when one considers the polymerization patterns of polyphenols and the food matrix in which they are inserted [17].

Many methods concentrate on the extraction of polyphenols since they are substances abundant in foods and effective antioxidants. Usually, these methods are based on the use of solvents with different polarities and in the study of the binomial time and temperature. The final yield of the extraction will depend on the solubility of the phenolics in the solvent used, the degree of polymerization, and its interactions with other constituents of food [18]. Free polyphenols can be extracted with water or with polar solvents such as methanol and acetone. However, some polymers may require the use of acids or enzymes, such as glycosidase [17]. Moreover, the use of long times and very high temperatures interferes negatively in the yield of the polyphenols [19]. The opposite may occur, depending on the polyphenol type and the matrix studied.

In this perspective, the use of response surface methodology becomes a useful and valuable tool in the field of antioxidant extraction. The factorial planning allows the study of the interaction of the various factors that interfere in the extraction of the polyphenols, minimizing costs and maximizing extraction yield. The advantages still include reducing of the number of experiments/replicates, simultaneous analysis of results, optimization of more than one response, and calculation of experimental error [20].

Thus, the present study was designed to determine the optimum condition for maximizing polyphenolic antioxidant extraction from fruits of *Byrsonima crassifolia* (L.) Kunth). As far as we know, this is the first investigation concerning the effects of temperature, time, solvent, and solid-liquid ratio on extraction of murici antioxidants.

2. Material and methods

2.1 Plant materials

A total of 10 kg of fresh fruits of *Byrsonima crassifolia* (L.) Kunth were purchased in a single batch from a rural producer in Araioses, Maranhão, Brazil. We used 10 kg of starting material to obtain representative amounts of fresh fruits. Since 50% of the fruits are composed by pits, the final weight of murici was 5 kg. The fruits were selected, washed, and kept frozen at −20°C until analysis.

2.2 Extraction procedures

For preparation of the extracts, the murici was weighed and homogenized with the solvent in turrax for 1 minute. The mixture was then subjected to magnetic stirring for 1 hour at 25°C. Subsequently, the blend was centrifuged at 15,000 G for 15 minutes at 4°C. The supernatant was collected and the volume completed to 50 mL.

2.3 Experimental design

The choice of solvent was the first step in the extraction of antioxidants. The most commonly used solvents are methanol, acetone, and mixtures thereof with water. So, initially, pretests were performed with the water and solvent blends: ethanol: water (60:40), methanol: water (60:40), and acetone: water (60:40). These experiments were carried out in order to verify the mixture with greater capacity for extraction of antioxidant compounds and help to decide which solvent should be used to optimize the extraction process.

In this pilot test, acetone in water (60%) provided the best extracting of antioxidants from murici (27.94 ± 0.28 mg polyphenols/100 mg extract and $IC50_{DPPH}$ 169.39 ± 10.62 μg/mL) and, thus, it was chosen as solvent in this study, while distilled water was the most inefficient solvent in the extraction of these compounds in murici (8.79 ± 0.09 mg polyphenols/100 mg extract and $IC50_{DPPH}$ 496.27 ± 8.53 μg/mL).

Response surface methodology (RSM) was employed to determine the optimum levels of the four in dependent variables (X1, solid-liquid ratio; X2, extraction time; X3, extraction temperature; and X4, solvent concentration), and five levels were used to evaluate the optimum combinations for antioxidant extraction. For the optimization of extraction, parameter designs consisting of 27 experiments, including three replicates in a central point, were used by the authors of [20]. The dependent variables studied were total phenolic content (TPC) and DPPH radical scavenging activity. The experimental runs for RSM were shown in **Table 1**. The variation ranges for the independent variables were established based on data from the literature and previous tests.

Independent variable			**Factor levels**		
	−2	−1	**0**	**+1**	**+2**
Solid-liquid ratio (mg/mL)	—	20	60	100	140
Time (minutes)	0	40	80	120	160
Temperature (°C)	10	25	40	55	70
Independent variable			**Factor levels**		
	−1.33	−1	**0**	**+1**	**1.66**
Solvent % (acetone in water)	10	20	50	80	100

Table 1.
Experimental domain of RSM.

2.4 Determination of total phenolic content (TPC)

The total phenolic compound (TPC) measurements in extracts were adapted from Singleton and Rossi [21] using gallic acid as a standard. Absorbance was measured at 750 nm. Results were expressed as mg equivalent gallic acid (GAE) per 100 mg of murici's dry weight (DW) (mg GAE/100 mg DW). All experiments were performed in three replicates.

2.5 DPPH radical scavenging activity

The extract (20 μL) was incubated with 150 μmol/L DPPH (200 μL) in 95% ethanol. Then, it was mixed and kept at room temperature in the dark for 30 min. Finally, the absorbance was measured at 515 nm using a microplate reader. The blank of each concentration was performed with ethanol instead of DPPH solution. The standard curve was a logarithm between 2 and 25 mg/mL gallic acid [22].

The % inhibition was calculated as described by the equation below. The antioxidant activity was expressed as IC50 (required extract concentration to reduce 50% of DPPH).

$$\%\ Inhibition = \frac{Absorbance\ Control - Absorbance\ Sample}{Absorbance\ Control} \times 100 \quad (1)$$

2.6 Statistical analysis

Statistical analysis was performed using the Statistica Software v.11 and fitted to a second-order polynomial regression model. An analysis of variance (ANOVA) with 95% confidence level was carried. The coefficient of determination (R^2) was used to estimate the fitness of the polynomial equation to the responses. The experimental and predicted values were compared in order to determine the validity of the model.

3. Results and discussion

The selection of solvent is as crucial stage when antioxidants extractions are concerned [23]. In order to evaluate the better solvent for the extraction of murici antioxidants, in terms of TPC and antioxidant activity (DPPH), a preliminary experiment was performed using ethanol, acetone, methanol, and water.

With the polarity change of the solvent, the amount of dissolved antioxidant compound varied. Solvents with low viscosity have low density and high diffusivity, which allows them to easily diffuse into the pores of vegetable materials, and consequently, extract the bioactive compounds [24]. As the result of that, the antioxidant activity of the extract also tends to increase, as observed in this study with the acetone. Hence, this solvent was deemed the better option in this research.

These results match with the results previously described where the mixture water:acetone is the more efficient at the extraction of phenolic antioxidants of oat grains [25] and *Etlingera elatior* Jack flowers [24], respectively. In addition, this mixture is also recommend as the best system for the extraction of polyphenols [26, 27].

In this study, RSM was employed to determine the optimum conditions for the extraction of antioxidant from fruits of *Byrsonima crassifolia* (L) Kunth (murici). In order to optimize the extraction parameters (extraction time, temperature, acetone concentration, and solid-liquid ratio), a five-level, four-factor experimental design was performed. The experimental conditions and results of 27 runs are presented in **Table 2**.

Original variables					TPC	IC_{50}
Test*	S/S	T (°C)	Time (min)	Acet. (%)		
1	20	25	40	20	24.57	146.12
2	100	25	40	20	24.38	144.06
3	20	55	40	20	23.90	162.26
4	100	55	40	20	23.55	165.41
5	20	25	120	20	23.15	152.07
6	100	25	120	20	23.27	157.85
7	20	55	120	20	20.79	174.72
8	100	55	120	20	20.86	179.86
9	20	25	40	80	21.39	160.05
10	100	25	40	80	21.73	162.49
11	20	55	40	80	17.50	188.85
12	100	55	40	80	18.06	184.94
13	20	25	120	80	19.84	163.47
14	100	25	120	80	20.07	169.16
15	20	55	120	80	12.83	205.09
16	100	55	120	80	13.21	203.07
17	60	40	80	50	27.03	132.73
18	60	40	80	50	27.52	128.68
19	60	40	80	50	26.58	137.29
20	20	40	80	50	27.80	129.90
21	140	40	80	50	24.51	142.48
22	60	10	80	50	23.73	167.38
23	60	70	80	50	20.81	192.18
24	60	40	0	50	24.95	144.93
25	60	40	160	50	20.29	186.91
26	60	40	80	10	16.58	201.93
27	60	40	80	100	13.73	218.58

**Does not correspond to order of processing. TPC—total phenolic content (mg GAE/100 mg), DPPH radical scavenging activity (IC_{50}—µg/mL), S/S—solid-solvent ratio (mg/mL), T—temperature, and Acet.—acetone in water.*

Table 2.
Response sheet for RSM experimental design with process variables and experimental results for the preparation of fruits of Byrsonima crassifolia *(L.) Kunth.*

The values of the TPC ranged from 13.21 to 27.80 mg GAE/100 mg, while the values of the IC 50 ranged from 128.68 to 218.58 µg/mL. The significance and model adequacy were tested by ANOVA, and it revealed that the model was highly significant ($p < 0.05$) for TPC and DPPH. The lack of fit of each model was not significant ($p > 0.05$), indicating that the developed model adequately explains the relationship between the independent variables and responses (**Tables 3** and **4**).

The linear and quadratic effects of the acetone concentration, temperature, and time extraction demonstrated significant effects on TPC ($p < 0.01$), as well as the

Source	Sum of squares	Degrees of freedom	F_{value}	R^2
Regression	385.75	6	18.42	0.8466
Residues	69.9	20		
Lack of fit	69.46	18	17.54	
Pure error	0.44	2		
Total	455.65	26		

Table 3.
ANOVA for the regression of the second-degree polynomial for the TPC response.

Source	Sum of squares	Degrees of freedom	F_{value}	R^2
Regression	12693.05	5	15.85	0.7906
Residues	3362.56	21		
Lack of fit	3325.45	19	9.43	
Pure error	37.11	2		
Total	16055.61	26		

Table 4.
ANOVA for the regression of the second-degree polynomial for the DPPH response.

interactions' temperature/time and time/acetone % ($p < 0.05$). In the case of DPPH, the linear and quadratic effects of temperature, time, and acetone % were significant ($p < 0.05$). The solid-liquid, in the studied range, did not interfere in the two responses ($p > 0.05$). Conversely, in the study of [28], the solid-to-solvent ratio (10–30 g/mL) showed a significant effect on the antioxidant compound extraction from germinated chickpea. Probably, an increase in the solvent to sample ratio helps the solute dissolve in solvent and hence the phenolic compounds extraction rate increased with decreasing of solid-to-liquid ratio [29]. However, the extraction rate trend stabilizes when the extraction rate reached a certain value [30]. In this study, we tested low solid-to-solvent ratios (20–140 mg/mL), which seems to have already provided maximum dissolution of the antioxidant compounds.

Analysis of the model clearly shows that temperature and acetone were the most significant factors. The acetone and temperature had a negative effect (coefficient = −4.98 and −3.46, respectively), indicating that a reduction in the temperature and acetone favored the recovery of TPC in the extract. A decrease in the acetone concentration enhanced the solvent polarity, which helps TPC dissolve in solvent [29]. Likewise, the acetone and temperature had a positive effect on DPPH-IC50 (coefficient = 19.35 and 26.12, respectively). This means that the extraction of antioxidants was less favorable at high temperature and acetone because a higher IC50 value corresponds to weaker antioxidant activity of tested extract. These results are similar to those found for [31] for mulberry (*Morus nigra*) pulp. On the other hand, high ethanol concentration and high temperature were the most effective factors for increasing TPC and antioxidative activities in mango peels [30].

As shown in **Figure 1**, the TPC increases rapidly as a result of acetone concentration, time and the temperature increase in the range of 10 and 45%, 40–50 min and 10 and 30°C, respectively. However, beyond acetone concentration of 50%, 55 min, and 35°C, the TPC decreases. Generally, temperature ranging from 60 to 70°C might increase the TPC yield due to higher solubility and diffusion in the solvent [30]. However, these temperatures are not recommended for polyphenol extraction from murici. It might be possible that higher temperatures for extended extraction caused the degradation of the some thermosensitive phenolic compounds

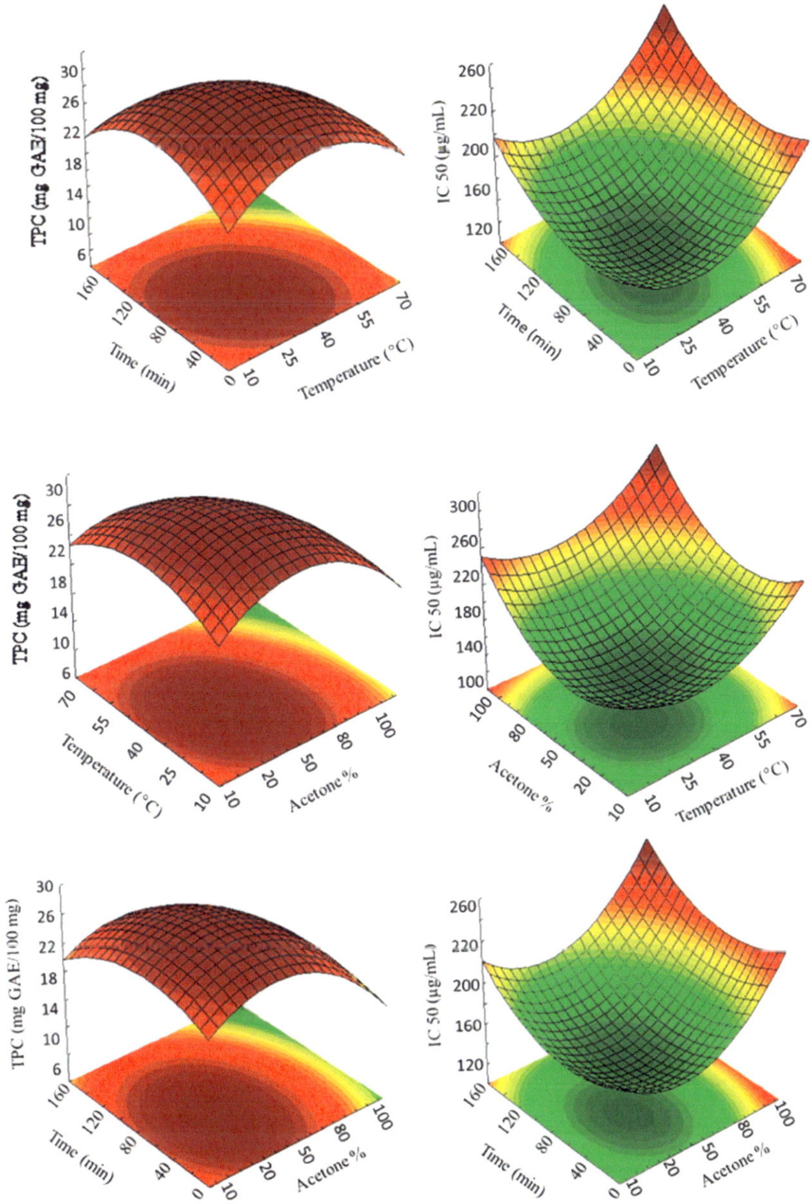

Figure 1.
Response surface plots for the effect of independent variables (stepping time, temperature, and acetone %) on total phenolic content (TPC) and total DPPH radical scavenging activity (IC50 - µg/mL) of fruits from Byrsonima crassifolia *(L.) Kunth.*

presented in murici [31, 15]. Similarly, the rise in acetone concentration, time, and the temperature decreased the antioxidant activity of murici after DPPH scavenging measurements. This result is in agreement with the findings to *Lycium ruthenicum* Murr. fruit [29] where lower solvent concentration was found to be conducive to total antioxidant activity.

The determined coefficient for TPC showed a good regression value (R^2 = 0.84), and the relationship between the TPC and the extraction parameters of the acetone concentration, the temperature, and the extraction time. By applying multiple regression analysis on the experimental data, the model for the response variable could be expressed by the following quadratic polynomial equation in the form of

	TPC (mg GAE/100 mg)	DPPH (IC_{50} - µg/mL)
Predicted value	27.61	137.65
Experimental value	28.04 ± 1.14*	139.00 ± 9.89*

**Mean ± standard deviation (n = 3 replicates).*

Table 5.
Predicted and experimental values under optimum conditions based on the multiple responses of TPC and DPPH.

coded values: $-\hat{y} = 26.42-1.40X_2-0.95X_2^2-1.27X_3-0.86X_3^2-1.87X_4-4.23X_4^2$, where X_2 = temperature, X_3 = time, and X_4 = acetone %.

The determined coefficient for DPPH also showed a reasonable regression value ($R^2 = 0.84$). The model for the response variable could be expressed by the following quadratic polynomial equation in the form of coded values:
$-\hat{y} = 141.9 + 10.77X_2 + 7.88X_2^2 + 7.29X_3 + 7.32X_4 + 23.35X_4^2$, where X_2 = temperature, X_3 = time, and X_4 = acetone %.

Based on the experimental results and statistical analysis, numerical optimizations were performed in order to determine the optimum level of independent variables. In our work, the optimum conditions for TPC compounds were as follows: extraction solvent using 44% acetone in water, an extraction temperature of 29°C, and an extraction time of 51 minutes. The optimum conditions for antioxidant activity (DPPH) were as follows: extraction solvent with 45% acetone in water, an extraction temperature of 40°C, and an extraction time of 53 minutes. These results suggest that the optimum parameters of extraction of murici's antioxidants occur with moderate percentages of acetone in water. The optimum temperature for antioxidants extraction evaluated by DPPH was in the middle range, and it was slightly higher for total phenolic compounds.

In order to verify the suitability of the response surface methodology model for quantitative predictions, experiments on estimated optimal conditions were performed. **Table 5** shows the predicted and experimental results for the variables selected. The predicted results matched well with the experimental results obtained at optimal extraction conditions proving the validity of the model to describe the process. These findings justified the selection of the RSM design, which was demonstrated to be accurate and reliable for predicting the TPC and antioxidant capacity of extracts from murici.

4. Conclusion

In this study, the RSM was successfully used to determine optimal levels of experimental parameters for the extraction of murici antioxidants. It is possible to verify that temperature, time, and acetone concentration negatively interfere in the extraction of polyphenols and in the antioxidant activity of fruits of *Byrsonima crassifolia* (L.) Kunth. The efficiency of the extraction can significantly be improved by using the novel extraction techniques, which may contribute to large-scale industrial applications and future pharmacological research.

Conflict of interest

All other authors report no potential conflict of interests.

Author details

Mariana Séfora Bezerra Sousa[1]*, Jovan Marques Lara Júnior[1] and Diego de Souza Buarque[2]

1 Food Department, Federal Institute of Piauí (IFPI), Brazil

2 Academic Unit of Serra Talhada (UAST), Rural Federal University of Pernambuco (UFRPE), Brazil

*Address all correspondence to: marianasefora@yahoo.com.br

References

[1] Vieira RF, Agostini-Costa TDS, Silva DD, Sano SM. FERREIRA FR. Frutas nativas da região centro-oeste do Brasil. 1ª ed. Vol. 1. Brasília: Distrito Federal; 2010. 322p. DOI: 10.13140/2.1.2141.120

[2] Lorenzi H. Árvores Brasileiras: Manual de Identificaçao e Cultivo de Plantas Arbóreas Nativas do Brasil. Vol. 1. São Paulo: Nova Odessa; 2009. 384p

[3] Silva MR et al. Caracterização química de frutos nativos do cerrado. Ciência Rural. 2008;**38**(6):1790-1793

[4] Barreto GPM, Benassi MT, Mercadante AZ. Bioactive compounds from several tropical fruits and correlation by multivariate analysis to free radical scavenger activity. Journal of the Brazilian Chemical Society. 2009; **20**(10):1856-1861. DOI: 10.1590/S0103-50532009001000013

[5] Godon A, Jungfer E, da Silva BA, Maia JGS, Marx F. Phenolic constituents and antioxidant capacity of four underutilized fruits from the Amazon region. Journal of Agricultural and Food Chemistry. 2011;**59**(14):7688-7699. DOI: 10.1021/jf201039r

[6] Mariutti LRB, Rodrigues E, Mercadante AZ. Carotenoids from *Byrsonima crassifolia*: Identification, quantification and in vitro scavenging capacity against peroxyl radicals. Journal of Food Composition and Analysis. 2013;**31**:155-160. DOI: 10.1016/j.jfca.2013.05.005

[7] Mariutti LR, Rodrigues E, Chisté RC, Fernandes E, Mercadante AZ. The Amazonian fruit *Byrsonima crassifolia* effectively scavenges reactive oxygen and nitrogen species and protects human erythrocytes against oxidative damage. Food Research International. 2014;**64**:618-625. DOI: 10.1016/j.foodres.2014.07.032

[8] de Souza VR, Pereira PAP, Queiroz F, Borges SV, Carneiro JDDS. Determination of bioactive compounds, antioxidant activity and chemical composition of Cerrado Brazilian fruits. Food Chemistry. 2012;**134**: 381-386. DOI: 10.1016/j.foodchem.2012.02.191

[9] Ramirez AM, Cotera LBF, Gutierrez RMZ. Anti-inflammatory activity of the hexane extract of *Byrsonima crassifolia* seeds in experimental animal models. Alternative Therapies in Health and Medicine. 2013;**19**(1):26-36

[10] Sousa MSB, Holanda IMSD, Monteiro HMC, Amâncio-dos-Santos Â. Antioxidant extract counteracts the effects of aging on cortical spreading depression and oxidative stress in the brain cortex. Acta Cirúrgica Brasileira. 2018;**33**(6):472-482. DOI: 10.1590/s0102-865020180060000001

[11] Perez-Gutierrez RM, Muñiz-Ramirez A, Gomez YG, Ramírez EB. Antihyperglycemic, antihyperlipidemic and antiglycation effects of *Byrsonima crassifolia* fruit and seed in normal and streptozotocin-induced diabetic rats. Plant Foods for Human Nutrition. 2010; **65**(4):350-357. DOI: 10.1007/s11130-010-0181-5

[12] Liu RH. Health benefits of fruit and vegetables are from additive and synergistic combinations of phytochemicals. American Journal of Clinical Nutrition. 2003;**78**(3):517S-520S. DOI: 10.1093/ajcn/78.3.517S

[13] Scalbert A, Johnson IT, Saltmarsh M. Polyphenols: Antioxidants and beyond. American Journal of Clinical Nutrition. 2005;**81**(1 Suppl):215S-217S. DOI: 10.1093/ajcn/81.1.215S

[14] Bravo L. Polyphenols: Chemistry, dietary sources, metabolism, and nutritional significance. Nutrition

Reviews. 1998;**56**(11):317-333. DOI: 10.1111/j.1753-4887.1998.tb01670.x

[15] Tomšik A, Pavlić B, Vladić J, Ramić M, Brindza VS. Optimization of ultrasound assisted extraction of bioactive compounds from wild garlic (*Allium ursinum* L.). Ultrasonics Sonochemistry. 2016;**29**:502-511. DOI: 10.1016/j.ultsonch.2015.11.005

[16] Angelo PM, Jorge N. Compostos Fenólicos em Alimentos: Uma breve revisão. Revista do Instituto Adolfo Lutz. 2007;**66**(1):232-240

[17] Tsao R. Chemistry and biochemistry of dietary polyphenols. Nutrients. 2010; **2**:1231-1246. DOI: 10.3390/nu2121231

[18] Andreo D, Jorge N. Antioxidantes naturais: Técnicas de extração. Boletim CEEPA. 2006;**24**(2):319-336. DOI: 10.5380/cep.v24i2.7489

[19] Moure A, Cruz JM, Franco D, Domınguez JM, Sineiro J, Domınguez H, et al. Natural antioxidants from residual sources. Food Chemistry. 2001;**72**: 145-171. DOI: 10.1016/S0308-8146(00) 00223-5

[20] Rodrigues MI, Iemma AF. Planejamento de Experimentos e Otimização de Processos: Uma estratégia sequencial de planejamentos. 1st ed. São Paulo: Campinas; 2005. 325p

[21] Singleton VL, Rossi JA Jr. Colorimetry of total phenolics with phosphomolybdic-phosphotungstic acid reagents. American Journal of Enology and Viticulture. 1965;**16**:144-158

[22] Brand-Williams W, Cuvelier M, Berset C. Use of a free radical method to evaluate antioxidant activity. Lebensmittel-Wissenschaft und Technologie. 1995;**28**(1):25-30. DOI: 10.1016/S0023-6438(95)80008-5

[23] Ali A, Lim XY, Chong CH, Mah SH, Chua BL. Optimization of ultrasound-assisted extraction of natural antioxidants from Piper betle using response surface methodology. LWT - Food Science and Technology. 2018; **89**:681-688. DOI: 10.1016/j.lwt.2017.11.033

[24] Wijekoon MMJO, Bhat R, Karim AA. Effect of extraction solvents on the phenolic compounds and antioxidant activities of bunga kantan (*Etlingera elatior*Jack.) inflorescence. Journal of Food Composition and Analysis. 2011; **24**(4–5):615-619. DOI: 10.1016/j.jfca.2010.09.018

[25] Zhou K, Yu L. Effects of extraction solvent on wheat bran antioxidant activity estimation. LWT - Food Science and Technology. 2004;**37**(7):717-721. DOI: 10.1016/j.lwt.2004.02.008

[26] Sulaiman SF, Sajak AAB, Ooi KL, Seow EM. Effect of solvents in extracting polyphenols and antioxidants of selected raw vegetables. Journal of Food Composition and Analysis. 2011; **24**(4–5):506-515. DOI: 10.1016/j.jfca.2011.01.020

[27] Michiels JA, Kevers C, Pincemail J, Defraigne JO, Dommes J. Extraction conditions can greatly influence antioxidant capacity assays in plant food matrices. Food Chemistry. 2012;**130**(4): 986-993. DOI: 10.1016/j.foodchem.2011.07.117

[28] Hayta M, Isçimen EM. Optimization of ultrasound-assisted antioxidant compounds extraction from germinated chickpea using response surface methodology. LWT - Food Science and Technology. 2017;**77**:208-216. DOI: 10.1016/j.lwt.2016.11.037

[29] Chen S, Zeng Z, Hu N, Bai B, Wang H, Suo Y. Simultaneous optimization of the ultrasound-assisted extraction for phenolic compounds content and antioxidant activity of Lycium ruthenicum Murr. Fruit using response surface methodology. Food Chemistry.

2018;**242**:1-8. DOI: 10.1016/j.
foodchem.2017.08.105

[30] Sai-Ut S, Benjakul S, Kraithong S, Rawdkuen S. Optimization of antioxidants and tyrosinase inhibitory activity in mango peels using response surface methodology. LWT - Food Science and Technology. 2015;**64**: 742-749. DOI: 10.1016/j.lwt.2015.06.003

[31] Espada-Bellido E, Ferreiro-González M, Carrera C, Palma M, Barroso CG, Barbero GF. Optimization of the ultrasound-assisted extraction of anthocyanins and total phenolic compounds in mulberry (*Morus nigra*) pulp. Food Chemistry. 2017;**219**:23-32. DOI: 10.1016/j.foodchem.2016.09.122

Section 2

Biosynthetic Routes of Phenolic Compounds

Chapter 3

Shikimic Acid Pathway in Biosynthesis of Phenolic Compounds

Norma Francenia Santos-Sánchez, Raúl Salas-Coronado, Beatriz Hernández-Carlos and Claudia Villanueva-Cañongo

Abstract

Phenolic compounds are secondary metabolites found most abundantly in plants. These aromatic molecules have important roles, as pigments, antioxidants, signaling agents, the structural element lignan, and as a defense mechanism. The expression of phenolic compounds is promoted by biotic and abiotic stresses (e.g., herbivores, pathogens, unfavorable temperature and pH, saline stress, heavy metal stress, and UVB and UVA radiation). These compounds are formed via the shikimate pathway in higher plants and microorganisms. The enzymes responsible for the regulation of phenolic metabolism are known, and shikimic acid is a central metabolite. The shikimate pathway consists of seven reaction steps, beginning with an aldol-type condensation of phosphoenolpyruvic acid (PEP) from the glycolytic pathway, and D-erythrose-4-phosphate, from the pentose phosphate cycle, to produce 3-deoxy-D-*arabino*-heptulosonic acid 7-phosphate (DAHP). A key branch-point compound is chorismic acid, the final product of the shikimate pathway. The shikimate pathway is described in this chapter, as well as factors that induce the synthesis of phenolic compounds in plants. Some representative examples that show the effect of biotic and abiotic stress on the production of phenolic compounds in plants are discussed.

Keywords: shikimate pathway, phenolic compounds, biosynthetic routes, phenylpropanoid metabolism

1. Introduction

The secondary metabolism is a biosynthetic source of several interesting compounds useful to chemical, food, agronomic, cosmetics, and pharmaceutical industries. The secondary pathways are not necessary for the survival of individual cells but benefit the plant as a whole [1]. Another general characteristic of secondary metabolism is that found in a specific organism, or groups of organisms, and is an expression of the individuality of species [2]. The secondary metabolism provides chemical diversity to organic molecules with low molecular weight that are related by the respective pathways; such organic molecules are called secondary metabolites. The secondary metabolites are often less than 1% of the total carbon in plant molecules [3]. These organic molecules isolated from terrestrial plants are the most studied, and their syntheses have an important role in the protection against

 IntechOpen

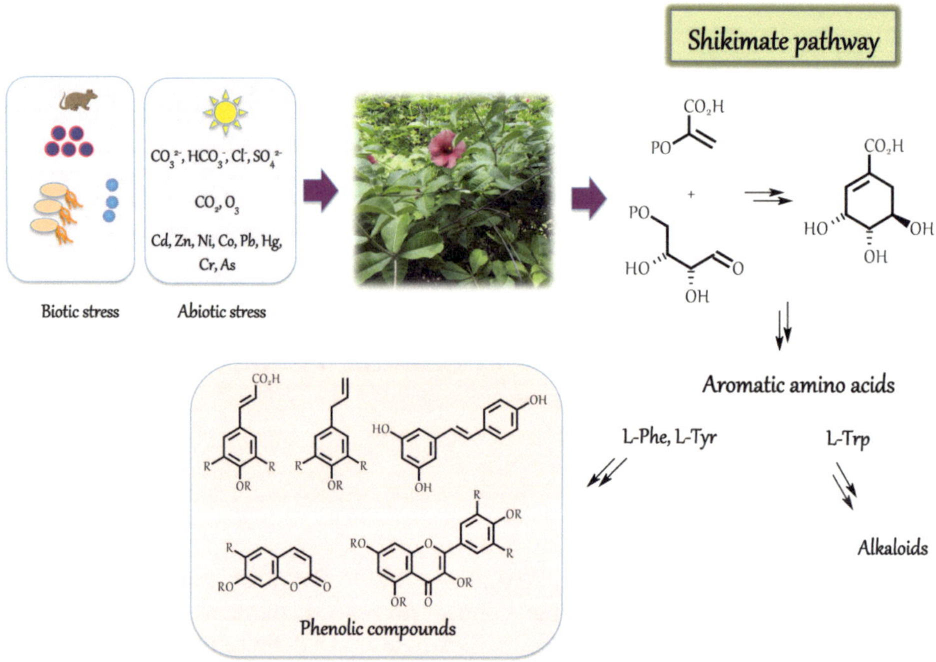

Figure 1.
Phenolic compound biosynthesis promoted by biotic and abiotic stresses (e.g., herbivores, pathogens, unfavorable temperature and pH, saline stress, CO_2, O_3, heavy metal stress, and UVB and UVA radiation).

pathogens, unfavorable temperature and pH, saline stress, heavy metal stress, and UVB and UVA radiation [3]. Secondary metabolism reflects plant environments more closely than primary metabolism [4]. There are three principal kinds of secondary metabolites biosynthesized by plants: phenolic compounds, terpenoids/isoprenoids, and alkaloids and glucosinolates (nitrogen- or sulfur-containing molecules, respectively) [5]. Phenolic compounds are biosynthesized by the shikimate pathway and are abundant in plants. The shikimate pathway, in plants, is localized in the chloroplast. These aromatic molecules have important roles, as pigments, antioxidants, signaling agents, electron transport, communication, the structural element lignan, and as a defense mechanism [6], **Figure 1**. The seven steps of the shikimate pathway and the metabolites for branch point are described in this chapter, as factors that induce the synthesis of phenolic compounds in plants. Some representative examples that show the effect of biotic and abiotic stress on the production of phenolic compounds in plants are discussed.

2. The shikimate pathway

The shikimate biosynthesis pathway provides precursors for aromatic molecules in bacteria, fungi, apicomplexan, and plants, but not in animals [2, 7]. Shikimic acid is named after the highly toxic Japanese *shikimi* (*Illicium anisatum*) flower from which it was first isolated [8]. This biochemical pathway is a major link between primary and secondary metabolism in higher plants [6]. In microorganisms, the shikimate pathway produces aromatic amino acids L-phenylalanine (L-Phe), L-tyrosine (L-Tyr), and L-tryptophan (L-Trp), molecular building blocks for protein biosynthesis [9]. But in plants, these aromatic amino acids are not only crucial components of protein biosynthesis; they also serve as precursors for diverse secondary metabolites that are important for plant growth [10]. These secondary metabolites are called

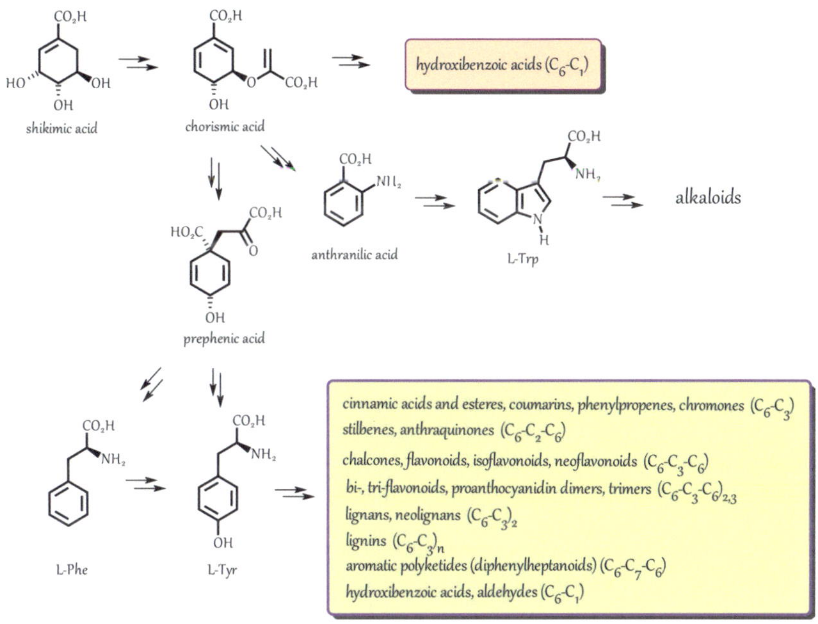

Figure 2.
The shikimic and chorismic acids are the common precursors for the synthesis of L-Phe, L-Tyr, and L-Trp and diverse phenolic compounds.

phenolic compounds and are synthesized when needed by the plant [11]. These molecules play an important role in the adaptation of plants to their ecosystem, and their study advances biochemical techniques and molecular biology [3, Bourgaud]. The principal aromatic phenolic compounds synthesized from L-Phe and L-Tyr are cinnamic acids and esters, coumarins, phenylpropenes, chromones (C_6-C_3), stilbenes, anthraquinones (C_6-C_2-C_6), chalcones, flavonoids, isoflavonoids, neoflavonoids (C_6-C_3-C_6), and their dimers and trimers, respectively $(C_6\ C_3\ C_6)_{2,3}$, lignans, neolignans $(C_6\text{-}C_3)_2$, lignans $(C_6\text{-}C_3)_n$, aromatic polyketides, or diphenylheptanoids (C_6-C_7-C_6) [12]. L-Trp is a precursor of alkaloids in the secondary metabolism [2]. Additionally, diverse hydroxybenzoic acids and aromatic aldehydes (C_6-C_1) are biosynthesized via branch points in the shikimate pathway, **Figure 2**. Phenolic compounds biosynthesized from the shikimate pathway have structural versatility.

The shikimate pathway consists of seven sequential enzymatic steps and begins with an aldol-type condensation of two phosphorylated active compounds, the phosphoenolpyruvic acid (PEP), from the glycolytic pathway, and the carbohydrate D-erythrose-4-phosphate, from the pentose phosphate cycle, to give 3-deoxy-D-*arabino*-heptulosonic acid 7-phosphate (DAHP), **Figure 3**. The seven enzymes that catalyze the pathway are known: 3-deoxy-D-*arabino*-heptulosonate-7-phosphate synthase (DAHPS; EC 4.1.2.15, now EC 2.5.1.54), 3-dehydroquinate synthase (DHQS; EC 4.2.3.4), 3-dehydroquinate dehydratase/shikimate dehydrogenase (DHQ/SDH; EC 4.2.1.10/EC 1.1.1.25), shikimate kinase (SK; EC 2.7.1.71), 5-enolpyruvylshikimate 3-phosphate synthase (EPSPS; EC 2.5.1.19), and chorismate synthase (CS; EC 4.2.3.5) [13], **Table 1**.

The shikimate pathway has special characteristics that are present only in bacteria, fungi, and plants. The absence of the pathway in all other organisms provides the enzymes catalyzing these reactions with potentially useful targets for the development of antibacterial agents and herbicides. For example, 5-*enol*pyruvylshikimate 3-phosphate synthase (EPSP-synthase) catalyzes the transfer of the enolpyruvyl (carboxyvinyl) moiety from PEP to shikimic acid 3-phosphate (S3P) [6].

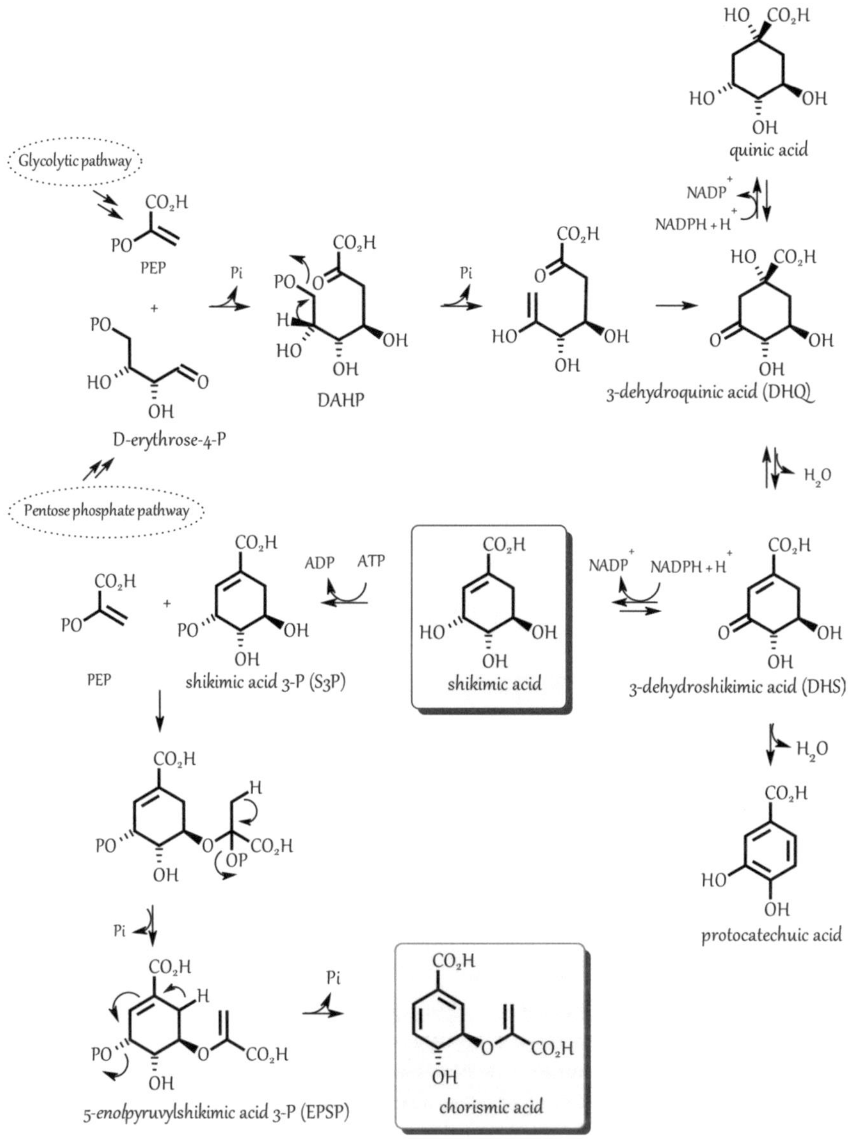

Figure 3.
Shikimate pathway.

In the second reaction step, DAHP loses phosphate (Pi); the enolic-type product is cyclized through a second aldol-type reaction to produce 3-dehydroquinic acid (DHQ). The 3-dehydroquinate synthase (DHQS) catalyzes this cyclization in the shikimate pathway. The DHQ dehydrates to produce 3-dehydroshikimic acid (DHS) (3-dehydroquinate dehydratase); this compound has a conjugated double carbon-carbon, **Figure 3**. The protocatechuic and the gallic acids (C_6-C_1) are produced by branch-point reactions from DHS [2]. The fourth step in the pathway is a reduction reaction of DHS with reduced nicotinamide adenine dinucleotide phosphate (NADPH), **Figure 3**. The fifth section of the pathway is the activation of shikimic acid with adenosine triphosphate (ATP) (shikimate kinase, SK) to make shikimic acid 3-phosphate (S3P). The sixth chemical reaction is the addition of PEP to S3P to generate 5-*enol*pyruvylshikimic acid 3-phosphate; the enzyme that catalyzes this reaction step, 5-*enol*pyruvylshikimate 3-phosphate synthase (EPSPS), has been extensively studied. The reason for this interest is because glyphosate [*N*-(phosphonomethyl)

Reaction step	Substrate	Enzyme/cofactor	Product
1	Phosphoenolpyruvate (PEP), erythrose-4-phosphate	3-Deoxy-D-*arabino*-heptulosonate-7-phosphate synthase (DAHPS; EC 4.1.2.15, now EC 2.5.1.54)/ Co^{2+}, Mg^{2+} or Mn^{2+} [15]	3-Deoxy-D-*arabino*-heptulosonic acid 7-phosphate (DAHP), Pi
2	3-Deoxy-D-*arabino*-heptulosonic acid 7-phosphate (DAHP)	3-Dehydroquinate synthase DHQS (EC. 4.2.3.4)/Co^{2+}, NAD^+ [15, 16]	3-Dehydroquinic acid (DHQ), Pi
3	3-Dehydroquinic acid (DHQ)	3-Dehydroquinate dehydratase (DHQ dehydratase EC 4.2.1.10) [15]	3-Dehydroshikimic acid (DHS), H_2O
4	3-Dehydroshikimic acid (DHS), NADPH + H^+	Shikimate dehydrogenase (SDH; EC 1.1.1.25) [18–21]	Shikimic acid, $NADP^+$
5	Shikimic acid, ATP	Shikimate kinase enzyme (SK; EC 2.7.1.71)	Shikimic acid 3-phosphate (S3P), ADP
6	Shikimic acid 3-phosphate (S3P), PEP	5-*Enol*pyruvylshikimate 3-phosphate synthase, also called aroA enzyme (EPSPS; EC 2.5.1.19) [25]	5-*Enol*pyruvylshikimate 3-phosphate (EPSP), Pi
7	5-*Enol*pyruvylshikimate 3-phosphate (EPSP)	Chorismate synthase (CS; EC 4.2.3.5)/$FMNH_2$ [2, 19, 30, 31]	Chorismic acid, Pi

Pi, phosphate; NAD^+, oxidized nicotinamide adenine dinucleotide; NADPH, reduced nicotinamide adenine dinucleotide phosphate; $FMNH_2$, reduced flavin mononucleotide.

Table 1.
Substrates, enzymes, and products of the shikimate pathway.

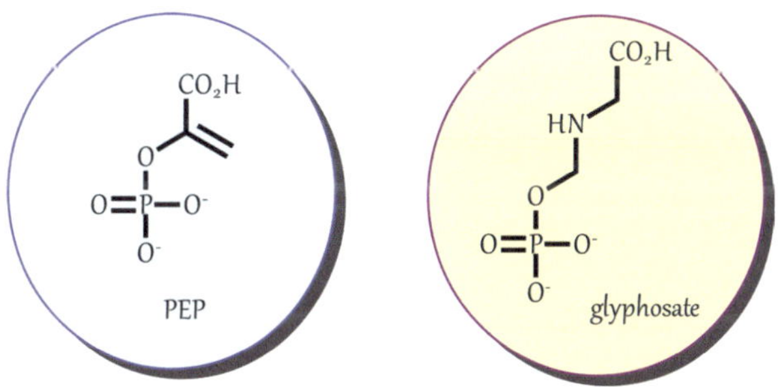

Figure 4.
PEP and glyphosate (powerful inhibitor of the 5-enolpyruvylshikimate 3-phosphate synthase, EPSPS).

glycine] is a powerful inhibitor of EPSPS [2], so glyphosate has been used as a broad-spectrum systemic herbicide. It is an organophosphorus molecule, phosphonic acid, and glycine derivative that has a similar molecular structure to PEP, **Figure 4**.

The last reaction step of the shikimate pathway is the production of chorismic acid from catalytic action on the chorismate synthase (CS). This reaction is a 1, 4-*trans* elimination of Pi, to yield the conjugated molecule, chorismic acid, **Figure 3**.

2.1 Synthesis of 3-deoxy-D-*arabino*-heptulosonic acid 7-phosphate (DAHP)

The first reaction of the shikimate pathway is an aldol-type condensation of PEP and carbohydrate erythrose-4-P, to give 3-deoxy-D-*arabino*-heptulosonic acid

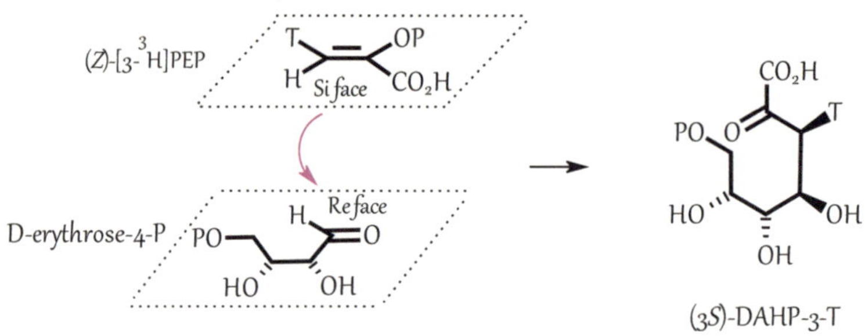

Figure 5.
Stereochemistry of the condensation reaction of (Z)-[3-^{3}H]PEP and D-erythrose-4-phosphate by DAHP synthase [14].

7-phosphate (DAHP), **Figures 3** and **5**. A new stereogenic center is generated in the condensation product DAHP catalyzed by the 3-deoxy-D-*arabino*-heptulosonate-7-phosphate synthase enzyme (DAHPS; EC 4.1.2.15, now EC 2.5.1.54). Results of enzymatic kinetic and labeled PEP with tritium (*Z*)-[3-^{3}H] PEP suggest that the nucleophilic attack of PEP is from the *Si* face of PEP to the *Re* face of the carbonyl group of D-erythrose-4-P, **Figure 5** [14]. Two isoenzymes of DAHPS have been found for the catalysis of this first reaction step. One isozyme needs only Mn^{2+}, and the other, either Co^{2+}, Mg^{2+}, or Mn^{2+} for the catalysis [15].

2.2 Synthesis of 3-dehydroquinic acid (DHQ)

The second reaction of the shikimate pathway is an intramolecular aldol-type reaction cyclization, where the enol (C6-C7) of DAHP nucleophilically attacks the carbonyl group (C2), to produce a six-member cycle, the 3-dehydroquinic acid (DHQ), **Figures 3** and 6. The enzyme that catalyzes this reaction, 3-dehydroquinate synthase DHQS (EC. 4.2.3.4), is a carbon-oxygen lyase enzyme that requires Co^{2+} and bound oxidized nicotinamide adenine dinucleotide (NAD^{+}) as cofactors [15, 16]. The Co^{2+} is essential for the catalytic activity of DHQS. Bender et al. [16] found that DHQS, from *Escherichia coli*, is a monomeric metalloenzyme that contains tightly bound Co^{2+}, and DHQS is deactivated with ethylenediaminetetraacetic acid (EDTA). The presence of the substrate (DAHP) blocks the inactivation by EDTA. The NAD^{+} cofactor dissociates form the DHQS enzyme rapidly in the presence of DAHP [16]. The reaction mechanism of the enzyme-catalyzed conversion of DAHP to DHQ involves five transformations from the DAHP hemiketal form, a pyranose: (1) oxidation of the hydroxyl at C5 adjacent to the lost proton that requires NAD^{+} (NAD^{+} need never dissociate from the active site), (2) the elimination of Pi of C7 to make the α,β-unsaturated ketone, (3) the reduction of C5 with NADH + H^{+}, (4) the ring opening of the enol to yield an enolate, and (5) the intramolecular aldol-like reaction to produce DHQ. All five-reaction steps occur through the function of DHQS, **Figure 6**.

The reduction reaction of DHQ leads to quinic acid at this branch point in the shikimate pathway. Quinic acid is a secondary metabolite that is free, forming esters or as part of alkaloids such as quinine. Quinic acid is found in high quantities in mature kiwi fruit (*Actinidia chinensis* and other species of *Actinidia*) and is a distinguishing characteristic of fresh kiwi fruit [7]. Also, the quinic acid is abundant in roasted coffee [17].

2.3 Synthesis of 3-dehydroshikimic acid (DHS) and shikimic acid

The third and fourth reaction steps of the shikimate pathway are catalyzed by a bifunctional enzyme: 3-dehydroquinate dehydratase/shikimate dehydrogenase

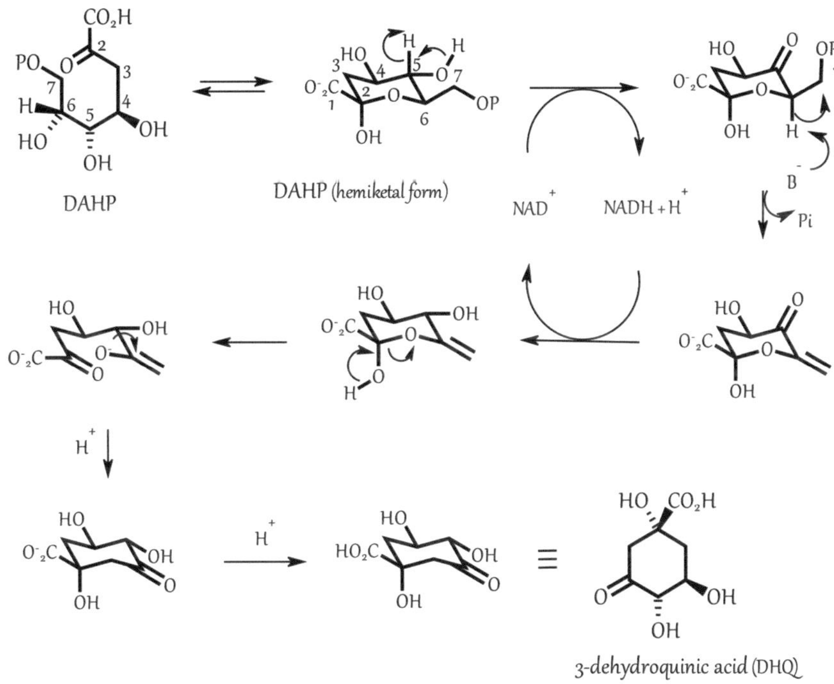

Figure 6.
Reaction mechanism of DAHP (hemiketal form) to 3-dehydroquinic acid (DHQ) by 3-dehydroquinate synthase DHQS (EC. 4.2.3.4) [16].

(DHQ dehydratase/SDH; EC 4.2.1.10/EC 1.1.1.25). The DHQ dehydratase enzyme is a hydro-lyase kind, and the SDH is an oxidoreductase enzyme. The DHQ dehydratase, in the third reaction step, converts DHQ into 3-dehydroshikimic acid (DHS) by eliminating water, and this reaction is reversible, **Figure 7**. The DHS is converted to shikimic acid in the fourth reaction step, by the reduction of the carbonyl group at C-5 by the catalytic action of SDH with NADPH, **Figure 3**. The biosynthesis of DHS is a branch point to shikimic acid and to the catabolic quinate pathway. If the DHS dehydrates, it produces protocatechuic acid (C_6-C_1) or gallic acid, **Figure 3**. Gallic acid (C_6-C_1) is a hydroxybenzoic acid that is a component of tannins [2].

Two structurally different kinds of 3-dehydroquinate dehydratase are known: type I (not heat-stable) and type II (heat-stable). Type I enzyme is present in bacteria and higher plants, and type II is found in fungi, which have both types of enzymes [18, 19]. The catalytic mechanism of the type I DHQ dehydratase has been detected by electrospray MS [20]. This catalytic mechanism involves the amino acid residue Lys-241 that forms a Schiff base with the substrate and product, **Figure 7** [21]. The fourth step is the reduction of DHS with NADPH that enantioselectively reduces the carbonyl of the ketone group of DHS to produce shikimic acid (shikimate dehydrogenase, SDH), **Figure 3**.

Sigh and Christendat [22] reported the crystal structure of DHQ dehydratase/SDH from the plant genus *Arabidopsis*. The crystal structure has the shikimate bound at the SDH and the tartrate molecule at the DHQ dehydratase. The studies show that Asp 423 and Lys 385 are key catalytic amino acids and Ser 336 is a key-binding group.

2.4 Synthesis of shikimic acid 3-phosphate (S3P)

The shikimate kinase enzyme (SK; EC 2.7.1.71) catalyzes the phosphorylation of the shikimic acid, the fifth chemical reaction of the shikimate pathway, and the products are shikimic acid 3-phosphate (S3P) and ADP, **Figures 3** and **8**.

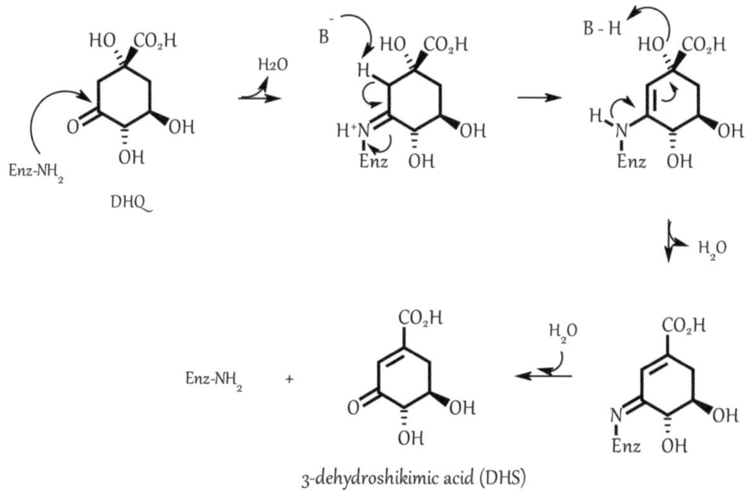

Figure 7.
Reaction mechanism to produce 3-dehydroshikimic acid (DHS) by type I DHQ dehydratase enzyme [21].

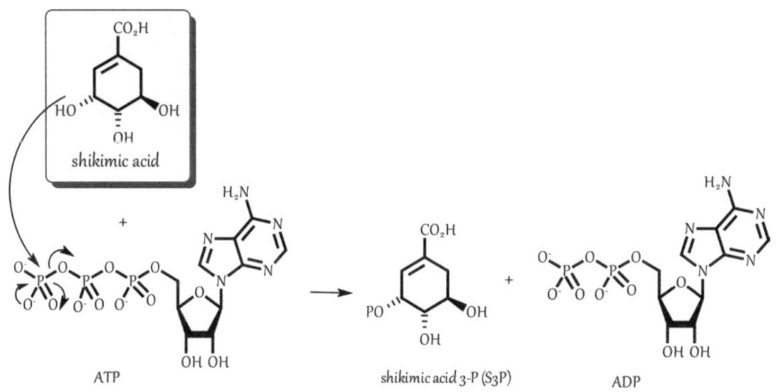

Figure 8.
Phosphorylation of shikimic acid with ATP.

Shikimic acid is phosphorylated with ATP in the 5-hydroxyl group of shikimic acid. SK is an essential enzyme in several bacterial pathogens and is not present in the human cell; therefore the SK enzyme has been classified as a protein target for drug design, especially for chemotherapeutic development of antitubercular drugs [23, 24].

2.5 Synthesis of 5-*enol*pyruvylshikimate 3-phosphate (EPSP)

The 5-*enol*pyruvylshikimate 3-phosphate synthase, also called aroA enzyme (EPSPS; EC 2.5.1.19), catalyzes the condensation of PEP to the 5-hydroxyl group of S3P in the sixth reaction of the shikimate pathway to form 5-*enol*pyruvylshikimate 3-phosphate (EPSP). The reaction mechanism involves the protonation of PEP to subsequent nucleophilic attack of the hydroxyl at C-5 of S3P to form an intermediate that loses Pi to form EPSP, **Figure 9** [25].

EPSPS is the most studied enzyme of the shikimate pathway because it plays a crucial role in the penultimate step. If this enzyme is inhibited, there is an

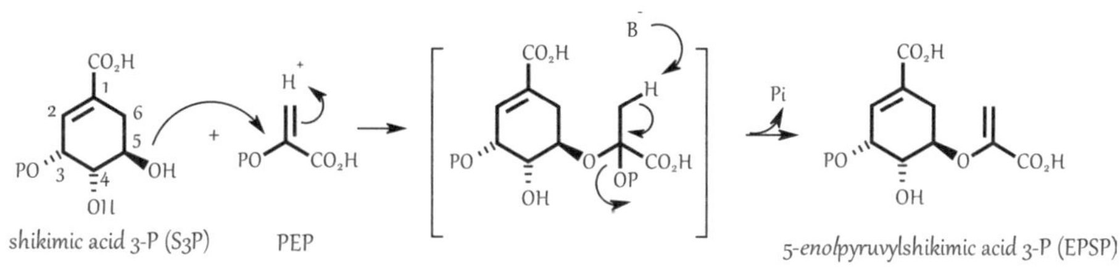

Figure 9.
Reaction mechanism of the condensation of S3P with PEP by EPSPS (EC 2.5.1.19) to form EPSP [25].

accumulation of shikimic acid [26], and the synthesis of aromatic amino acid is disabled, leading to the death of the plant [27]. Therefore, EPSPS is used as a target for pesticides, like glyphosate, **Figure 4**, the active ingredient in the herbicides RoundUp™, Monsanto Chemical Co., and Touchdown™, Syngenta. Glyphosate (*N*-(phosphonomethyl)glycine) inhibits EPSPS and is a potent nonselective herbicide that mimics the carbocation of PEP and binds EPEPS competitively [28]. Because the glyphosate is nonselective and kills food crops, there is interest in finding glyphosate-tolerant genes for genetically modified crops [29]. Two types of EPSPS enzymes have been identified: type I EPSPS (sensitive to glyphosate) identified mostly in plants and bacteria and type II EPSPS (nonsensitive to glyphosate and has a high affinity for PEP), found in some bacteria [27].

2.6 Synthesis of chorismic acid

The seventh and last reaction step of the shikimate pathway is the 1,4-*trans* elimination of the Pi group at C-3 from EPSPS to synthetize chorismic acid. This last step is catalyzed by chorismate synthase (CS; EC 4.2.3.5) that needs reduced flavin mononucleotide ($FMNH_2$) as a cofactor that is not consumed [2, 19]. The $FMNH_2$ transfers an electron to the substrate reversibly [30]. Spectroscopic

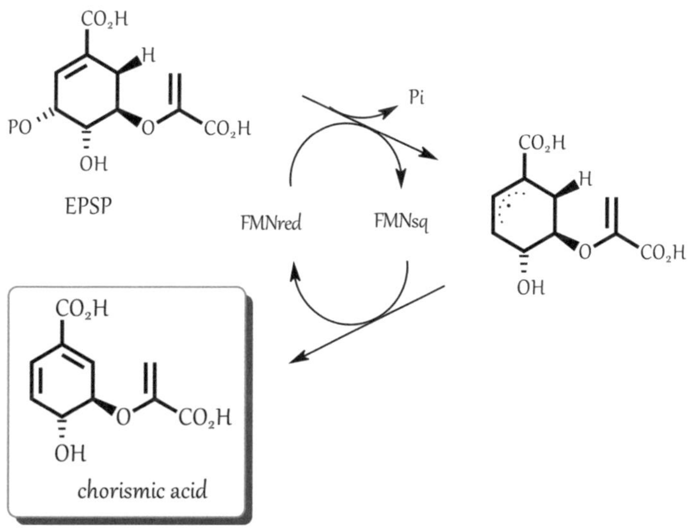

Figure 10.
Reaction of mechanism to yield chorismic acid by chorismate synthase [30].

techniques and kinetic isotope effect studies suggest that a radical intermediate in a non-concerted mechanism is developed [30, 31], **Figure 10**. Chorismic acid, the final molecule of the shikimate pathway, is a key branch point to post-chorismic acid pathways, to obtain L-Phe, L-Tyr, and L-Trp, **Figure 2**. L-Phe is the substrate to phenylpropanoid and flavonoid pathways [13].

3. Factors that induce the synthesis of phenolic compounds in plants

The expression of phenolic compounds is promoted by biotic and abiotic stresses (e.g., herbivores, pathogens, unfavorable temperature and pH, saline stress, heavy metal stress, and UVB and UVA radiation). UV radiation is divided into UVC (≤280 nm), UVB (280–320 nm), and UVA (300–400 nm). UVA and UVB radiation are transmitted through the atmosphere; all UVC and some UVB radiation (highly energetic) are absorbed by the Earth's ozone layer. This accumulation is explained by the increase in enzymatic activity of the phenylalanine ammonia-lyase and chalcone synthase enzymes, among others [12]. Studies have been done about the increase of phenolic compounds, such as anthocyanins, in plants when they are exposed to UVB radiation [13]. Another study demonstrates that UVB exposure enhances anthocyanin biosynthesis in "Cripps pink" apples (*Malus* x *domestica* Borkh.) but not in "Forelle" pears (*Pyrus communis* L.) [32]. This effect may be due to UV radiation exposure and the cultivar of the plants studied. It is known that if plants are under stress, they accumulate phenolic compounds.

The increase in phenolic compounds in blueberry (*Vaccinium corymbosum*) plantlets cultivated in vitro exposed to aluminum (Al) and cadmium (Cd) has also been studied. These heavy metals cause high toxicity in plants, because they increase the oxidative stress by the production of reactive oxygen species (ROS). The authors of the study suggest that the phenolic compounds, specifically chlorogenic and ellagic acids, **Figure 11**, reduce the ROS in blueberry plants [33].

An interesting study was carried out in 2011 by Mody et al., where they studied the effect of the resistance response of apple tree seedlings (*Malus* x *domestica*) to a leaf-chewing insect (*Spodoptera littoralis*) [34]. The authors found a significant herbivore preference for undamaged plants (induced resistance) was first observed 3 days after herbivore damage in the most apical leaf. Also, the results showed higher concentrations of the flavonoid phlorizin, **Figure 12**, in damaged plants than undamaged plants. This indicates that insect preference for undamaged apple plants may be linked to phlorizin, which is the main secondary metabolite of the phenolic type in apple leaves.

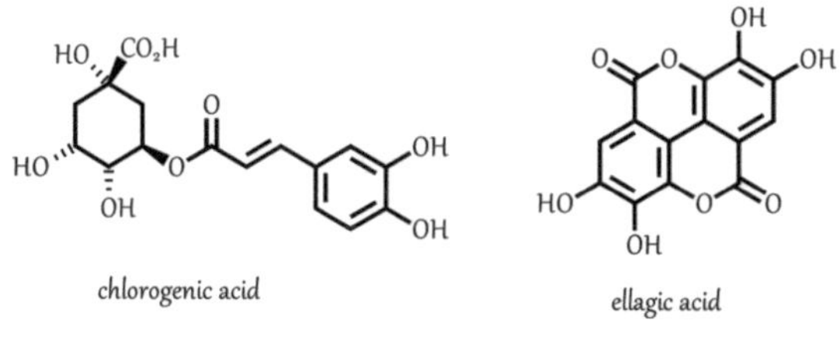

Figure 11.
Chemical structure of chlorogenic (C_6-C_3) and ellagic (C_6-C_1) acids.

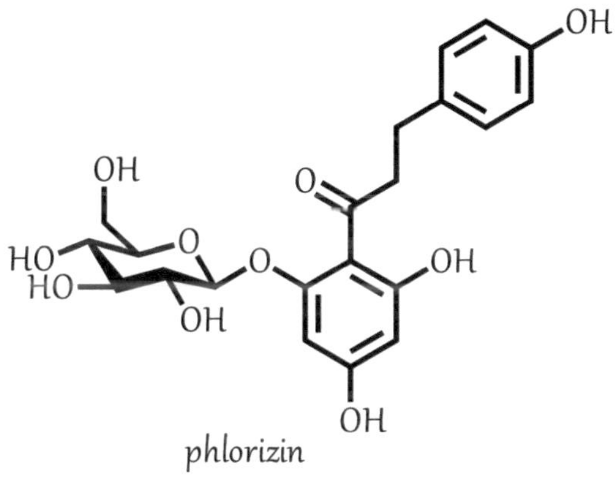

Figure 12.
Chemical structure of phlorizin (C_6-C_3).

4. Conclusions

Knowledge of the biosynthetic pathway of shikimic acid leads to understanding the reaction mechanisms of enzymes and thus discovering antimicrobials, pesticides, and antifungals. Studies with isotopic labeling of substrates, the use of X-ray diffraction, nuclear magnetic resonance (NMR), mass spectrometry (ES), biotechnology, as well as organic synthesis have contributed to explaining the shikimate pathway. Although the seven steps of the biosynthetic pathway are elucidated, these metabolites are the precursors of phenolic compounds, more complex molecules that are necessary for the adaptation of plants to the environment. So, the shikimate pathway is the basis for the subsequent biosynthesis of phenolic compounds. There is scientific interest in continuing to investigate the biosynthesis of phenolic compounds from several points of view: pharmaceuticals, agronomy, chemical and food industries, genetics, and health.

Acknowledgements

The authors thank Carol Ann Hayenga for her English assistance in the preparation of this manuscript. The Technological University of the Mixteca provided support.

Conflict of interest

The authors have no conflict of interest to declare and are responsible for the content and writing of the manuscript.

Ethical approval

This chapter does not contain any studies with human participants or animals performed by any of the authors.

Author details

Norma Francenia Santos-Sánchez*, Raúl Salas-Coronado,
Beatriz Hernández-Carlos and Claudia Villanueva-Cañongo
Institute of Agroindustry, Technological University of the Mixteca, Huajuapan de León, Oaxaca, Mexico

*Address all correspondence to: nsantos@mixteco.utm.mx

References

[1] Adams ZP, Ehlting J, Edwards R. The regulatory role of shikimate in plant phenylalanine metabolism. Journal of Theoretical Biology. 2019;**462**:158-170. DOI: 10.1016/j.jtbi.2018.11.005

[2] Dewick PM. Medicinal Natural Products: A Biosynthetic Approach. 3rd ed. United Kingdom: John Wiley and Sons Ltd.; 2009. p. 539. DOI: 10.1002/9780470742761

[3] Bourgaud F, Gravot A, Milesi S, Gontier E. Production of plant secondary metabolites: A historical perspective. Plant Science. 2001;**161**:839-851. DOI: 10.1016/S0168-9452(01)00490-3

[4] Yang D, Huang Z, Jin W, Xia P, Jia Q, Yang Z, et al. DNA methylation: A new regulator of phenolic acids biosynthesis in *Salvia miltiorrhiza*. Industrial Crops and Products. 2018;**124**:402-411. DOI: 10.1016/j.indcrop.2018.07046

[5] Aharoni A, Galili G. Metabolic engineering of the plant primary-secondary metabolism interface. Current Opinion in Biotechnology. 2011;**22**:239-244. DOI: 10.1016/j.copbio.2010.11.004

[6] Macheroux P, Schmid J, Amrhein N, Schaller A. A unique reaction in a common pathway: Mechanism and function of chorismate synthase in the shikimate pathway. Planta. 1999;**207**:325-334

[7] Mittelstädt G, Negron L, Schofiel LR, Marsh K, Parker EJ. Biochemical and structural characterisation of dehydroquinate synthase from the New Zealand kiwifruit *Actinidia chinensis*. Archives of Biochemistry and Biophysics. 2013;**537**:185-191. DOI: 10.1016/j.abb.2013.07.022

[8] Ghosh S, Chisti Y, Banerjee UC. Production of shikimic acid. Biotechnology Advances. 2012;**30**:1425-1431. DOI: 10.1016/j.biotechadv.2012.03.001

[9] Weaver LM, Herrmann KM. Dynamic of the shikimate pathway. Trends in Plant Science. 1997;**9**:346-351

[10] Tzin V, Galili G. Amino acids biosynthesis pathways in plants. Molecular Plant. 2010;**3**:956-972. DOI: 10.1093/mp/ssq048

[11] Dixon RA, Strack D. Phytochemistry meets genome analysis, and beyond. Phytochemistry. 2003;**62**:815-816

[12] Cheynier V, Comte G, Davies KM, Lattanzio V, Martens S. Plant phenolics: Recent advances on their biosynthesis, genetics and ecophysiology. Plant Physiology and Biochemistry. 2013;**72**:1-20. DOI: 10.1016/j.plaphy.2013.05.009

[13] Zhang Z-Z, Li X-X, Chu Y-N, Zhang M-X, Wen Y-Q, Duan C-Q, et al. Three types of ultraviolet irradiation differentially promote expression of shikimate pathway genes and production of anthocyanins in grape berries. Plant Physiology and Biochemistry. 2012;**57**:74-83. DOI: 10.1016/j.plaphy.2012.05.005

[14] Floss HG, Onderka DK, Carroll M. Stereochemistry of the 3-deoxy-D-arabino-heptulosonate 7-phosphate synthetase reaction and the chorismite synthetase reaction. The Journal of Biological Chemistry. 1972;**247**:736-744

[15] Schmid J, Amrhein N. Molecular organization of the shikimate pathway in higher plants. Phytochemistry. 1995;**39**:737-749

[16] Bender SL, Mehdi S, Knowles JR. Dehydroquinate synthase: The role of divalent metal cations and of nicotinamide adenine dinucleotide in catalysis. Biochemistry. 1989;**28**:7555-7560

[17] Scholz BM, Maier G. Isomers of quinic acid and quinide in roasted coffee. Zeitschrift für Lebensmittel-Untersuchung und -Forschung. 1990;**190**:132-134

[18] Harris JM, Gonzalez-Bello C, Kleanthous C, Hawkins A, Coggins J, Abell C. Evidence from kinetic isotope studies for an enolate intermediate in the mechanism of type II dehydroquinases. Biochemical Journal. 1999;**319**:333-336

[19] Hermann KM. The shikimate pathway: Early steps in the biosynthesis of aromatic compounds. The Plant Cell. 1995;7:907-919

[20] Shneier A, Kleanthous C, Deka R, Coggins JR, Abel C. Observation of an imine intermediate on dehydroquinase by electrospray mass spectrometry. Journal of the American Chemical Society. 1991;**113**:9416-9418

[21] Sigh SA, Christendat D. The DHQ-dehydroshikimate-SDH-shikimate-NADP(H) complex: Insights into metabolite transfer in the shikimate pathway. Crystal Growth & Design. 2007;7:2153-2160

[22] Sigh SA, Christendat D. Structure of *Arabidopsis* dehydroquinate dehydratase-shikimate dehydrogenase and implications for metabolic channeling in the shikimate pathway. Biochemistry. 2006;**45**:7787-7796. DOI: 10.1021/bi060366+

[23] Coracini JD, de Azevedo WF Jr. Shikimate kinase, a protein target for drug design. Current Medicinal Chemistry. 2014;**21**:592-604. DOI: 10.2174/09298673113206660299

[24] Blanco B, Prado V, Lence E, Otero JM, García-Doval C, van Raaij MJ, et al. *Mycobacterium tuberculosis* shikimate kinase inhibitors: Design and simulation studies of the catalytic turnover. Journal of the American Chemical Society. 2013;**135**:12366-12376. DOI: 10.1021/ja405853p

[25] Lewis J, Johnson KA, Anderson KS. The catalytic mechanism of EPSP synthase revisited. Biochemistry. 1999:7372-7379

[26] Maroli A, Nandula V, Duke S, Tharayil N. Stable isotope resolved metabolomics reveals the role of anabolic and catabolic processes in glyphosate-induced amino acid accumulation in *Amaranthus palmeri* biotypes. Journal of Agricultural and Food Chemistry. 2016;**64**:7040-7048. DOI: 10.1021/acs.jafc.6b02196

[27] Cao G, Liu Y, Zhang S, Yang X, Chen R, Zhang Y, et al. A novel 5-enolpyruvylshikimate-3-phosphate synthase shows high glyphosate tolerance in *Escherichia coli* and *Tobacco Plants*. PLoS One. 2012;7:e38718. DOI: 10.1371/journal.pone.0038718

[28] Yi S-y, Cui Y, Zhao Y, Z-d L, Y-j L, Zhou F. A novel naturally occurring class I 5-enolpyruvyl shikimate-3-phosphate synthase from *Janibacter* sp. confers high glyphosate tolerance to rice. Scientific Reports. 2016;**6**:1904

[29] Liu F, Cao Y. Expression of a bacterial aroA gene confers tolerance to glyphosate in tobacco plants. Turkish Journal of Biology. 2018;**42**:187-194. DOI: 10.3906/biy-1712-56

[30] Bornemann S, Theoclitou M-E, Brune M, Webb MR, Thorneley RNF, Abell C. A secondary β deuterium kinetic isotope effect in the chorismate synthase reaction. Bioorganic Chemistry. 2000;**28**:191-204. DOI: 10.1006/bioo.2000.1174

[31] Osborne A, Thorneley RNF, Abell C, Bornemann S. Studies with substrate and cofactor analogues provide evidence for radical mechanism in the chorismate synthase reaction.

The Journal of Biological Chemistry. 2000;**275**:35825-35830

[32] Marais E, Jacobs G, Holcroft DM. Postharvest irradiation enhances anthocyanin synthesis in apples but nor in pears. HortScience. 2001;**36**:738-740

[33] Manquián-Cerda K, Cruces E, Escudey M, Zúñiga G, Calderón R. Interactive effects of aluminum and cadmium on phenolic compounds, antioxidant enzyme activity and oxidative stress in blueberry (*Vaccinium corymbosum* L.) plantlets cultivated in vitro. Ecotoxicology and Environmental Safety. 2018;**150**:320-326. DOI: 10.1016/j.ecoenv.2017.12.050

[34] Gutbrod B, Mody K, Wittwer R, Dorn S. Within-plant distribution of induced resistance in apple seedlings: Rapid acropetal and delayed basipetal responses. Planta. 2011;**233**:1199-1207

Chapter 4

MicroRNAs Associated with Secondary Metabolites Production

Vargas-Hernández Marcela, Vázquez-Marrufo Gerardo, Aguilar-Ruiz Carlos Agustín, González-Márquez Marco Antonio, Rocha Oscar, Cerna-Pantoja Diego and Andrés Cruz-Hernández

Abstract

MicroRNAs (miRNAs) are noncoding RNAs that play an important role in the regulation of the genetic expression in animals and plants by targeting mRNAs for cleavage or translational repression. Several miRNAs regulate the plant development, the metabolism, and the responses to biotic and abiotic stresses. Characterization of an miRNA has helped to show its role in fine tuning the mechanisms of posttranscriptional gene regulation. Although there is a lot of information related to miRNA regulation of some processes, the role of miRNA involved in the regulation of biosynthesis of secondary plant product is still poorly understood. In this chapter, we summarize the identification and characterization of miRNAs that participate in the regulation of the biosynthesis of secondary metabolites in plants and their use in the strategies to manipulate a controlled manipulation.

Keywords: alkaloids, flavonoids, flavonols, gene expression, isoprenoids

1. Introduction

1.1 Definition of miRNA

mRNAs are noncoding single-stranded RNA molecules that range from a length of 18 to 28 nucleotides. These molecules play an important role in post-transcriptional regulation by the inhibition of the expression of target genes by binding to mRNA [1]. Eukaryotic organisms, such as plants and animals, and some viruses express this type of molecules [2, 3]. Lin-4 is the first miRNA that was identified in *Caenorhabditis elegans*, although later in the same organism were found 22-nt lin-4 and 21-nt let-7; currently, more than 18,226 miRNAs are reported [4]. Although miRNA:target-gene interactions are widely conserved, this process is limited between kingdoms [5]. miRNAs are distributed in genome as single or clusters expressed as polycistronic units and share function relationships [6]. In plants, most miRNAs are encoded by their own primary transcript; few examples of miRNA cluster are reported (i.e., miR395). Introns are the main hotspots for their origination [7]. Some of the mechanisms where they act are the development of time and host-pathogen interactions, as well as cell differentiation, proliferation, apoptosis, and tumorigenesis.

IntechOpen

1.2 Mechanism of action of miRNAs

The biogenesis of miRNAs in plants and animals presents some differences. In both plants and animals, the precursors of miRNAs are polyadenylated caps and RNAs, and transcribed as for any coding RNA by RNA polymerase II (RNAPII) [5, 8]. However, in the plants, the primary transcript (pri-miRNA) that gives rise to the miRNA is produced by the nuclear RNAase dicer-like 1 (DCL1) and its accessory proteins SERRATE (SE) and hyponastic leaves (HYL1) [8]. Also, Drosha gene is absent in plants [5].

In *Arabidopsis thaliana*, DCL1 and hyponastic leaves 1 (HYL1) cleave the pri-miRNA in the nucleus of the cell, which gives rise to the precursor-miRNA (pre-miRNA) dsRNA. Subsequently, there is another cleavage by the action of DCL1 and HYL1 to release the miRNA; the two nucleotide 3′ overhangs are methylated by the action of the sRNA-specific methyltransferase HUA enhancer1 (HEN1). When the mature single-stranded miRNA is found in the cytoplasm, it is loaded onto AGO1 that is present in RNA-induced silencing complexes (RISC), repressing the expression by mRNA cleavage [9]. 3′ untranslated region (3′ UTR) is the union region of the miRNAs to its target mRNAs which allows it to be repressed [10].

The expression of miRNAs is regulated by transcription factors. Negative on TATA less 2 (NOT2) promotes the transcription of protein miRNA genes and facilitates efficient DCL1 recruitment in miRNA biogenesis [11]. Cell division cycle 5 (CDC5) acts as a positive transcription factor associating with miRNA genes [12]. Pleiotropic regulatory locus 1 (PRL1) has the ability to bind to DCL1 and pri-miRNAs. The miRNA duplex is transported to the cytoplasm by nuclear export factor *Drosophila* Exportin-5 ortholog HASTY (HST).

The miRNA target genes can be a single member of a gene family or regulate a multiple family members. Thus, multiple miRNA genes could be targeting a single member, with tissues and stage specificity, and/or a single miRNA gene could be regulating multiple family members. The spatial and temporal expression and abundance of mature miRNAs are tightly regulated; they vary greatly among different miRNAs; and the abundance also varies depending on the tissue types or developmental stages [13].

1.3 Regulatory processes involving miRNAs

In the cytoplasm of cells, the miRNAs regulate the expression of genes at the posttranscriptional level via mRNA degradation and/or translational repression [14]. Unlike animals, in plants, there is a perfect complementarity between miRNA and target mRNA [14]. To carry out the silencing, a ribonucleoprotein RNA-induced silencing complex (RISC) is formed [15]. AGO1, AGO2, AGO4, AGO7, and AGO10 slicer activity has been reported, even though AGO1 is associated with most miRNAs [16]. AGO1-catalyzed RNA cleavage (slicing) represses miRNA targets [17].

2. miRNAs and secondary metabolism in plants

In plants, miRNAs control the expression of genes that encode transcription factors, stress response proteins, and others, which have an impact on biological processes. The miRNAs regulate the biological processes in the plans such as maintenance of genome integrity, primary and secondary metabolism, development, signal transduction, signaling pathways, homeostasis, innate immunity, and adaptive responses to biotic and abiotic stress [18]. Secondary metabolites are a group of phytochemicals that regulate various processes related to the interaction of the plant with its environment [19]. These compounds include terpenoids, alkaloids, phenolics, glycosides, tannins, and saponins, and defend plants from several biotic an abiotic stressors [20]. Even though these types of compounds are synthesized by plants to help in self-defense, they have diverse industrial uses such as insecticides, dyes, flavoring compounds, and nutraceuticals having a positive effect on human health. Commercial importance has resulted in a great interest in studying possibilities of enhancing its production [21]. It is known that miRNAs control several biological processes at the posttranscriptional level. Currently, some studies reveal the role that miRNAs have in the regulation of secondary metabolic pathways [20]. Therefore, the production of compounds derived from secondary metabolism can be managed through the miRNAs. Since they are positively or negatively regulated, the production of desired metabolites can be induced, the production of toxic metabolites can be limited, and new metabolites can be produced [22].

Computational analysis carried out in two transcriptomes of Swertia resulted in the identification of miRNAs associated to secondary metabolites biosynthesis; miR-156a, miR-166a, miR-166b, miR-168, miR-11071, and miR-11320 targeting metabolic enzymes, such as aspartate aminotransferase, ribulose-phosphate 3-epimerase, acetyl-CoA acetyltransferase, phosphoglycerate mutase, and premnaspirodiene oxygenase, also include a gene encoding a homeobox-leucine zipper protein (HD-ZIP) with a possible association in secondary metabolites biosynthesis in *Swertia chirayita* [23]. Elicited or infected plants induce change in gene expression and production of defensive metabolites and these might be regulated by miRNAs. *Solanum tuberosum* L. under light stimulus found light-responsive miRNAs that are important regulators in alkaloid metabolism, UMP salvage, lipid biosynthesis, and cellulose catabolism [24]. Cadmium stress in oilseed rape (*Brassica napus* L.) reported miRNAs involved in the regulations of TFs, biotic stress defense, ion homeostasis, and secondary metabolism synthesis [25]. *Nicotiana tabacum* plants infected with tobacco mosaic virus (TMV), at the early stage of infection (5 dpi), show a cluster of miRNAs with down-accumulation, while most of the miRNAs were upregulated at 15 and 22 dpi, including both miRNAs and miRNA targets [26].

2.1 Flavonoids

Flavonoids are secondary metabolites that possess a polyphenolic structure. Those compounds consist of hydroxylated phenolic substances having a benzo-γ-pyrone structure and derived of phenylpropanoid pathway [27]. Within the subgroups of the flavonoids are flavones, flavonols, flavanones, flavanonols, flavanols or catechins, anthocyanins, and chalcones [28]. For plants, this type of compounds is synthesized as a result of the interaction with the environment, other plants, and microorganisms. They have diverse biological functions as anti-oxidative, anti-inflammatory, anti-mutagenic, and anti-carcinogenic properties, which are structure dependent [28]. The above makes flavonoids a compound with nutraceutical, pharmaceutical, medicinal, and cosmetic applications [28]. The production of secondary metabolites is found in cases regulated by the miRNAs (**Table 1**). Little is known about the miRNAs involved in flavonoid biosynthesis. In *Helianthus*, 323,318 ESTs were computationally screened for the miRNAs identification of them, and a miR911 family was found related to the biosynthesis of tocopherols. Gou et al. [51] demonstrate that accumulation of anthocyanins in the stems of *A. thaliana* is under the regulation of miR156-targeted squamosa promoter binding protein-like (SPL) genes. High miR156 activity promotes accumulation of anthocyanins and activity-induced of flavonols. This study also demonstrates that SPL9 negatively regulates anthocyanin accumulation through

Plant species	miRNA	Target	Function	References
Flavonoids				
Sunflower	miR2911	Gamma-tocopherol methyl transferase	Tocopherols biosynthesis	[29]
A. thaliana	miR156	SPL transcription factor	Accumulation of anthocyanins, whereas reduced miR156 activity results in high levels of flavonols	[51]
Diospyros kaki	miR395p-3p and miR858b	bHLH and MYB, respectively	Proanthocyanidin biosynthesis	[30]
Lonicera japonica	miRNAs (U436803, U977315, U805963, U3938865 and U4351355)	R2R3-MYB transcription factors	Flavonoid biosynthesis	[2, 3]
A. thaliana	MicroRNA858a	R2R3-MYB transcription factors	Flavonoid biosynthesis	[31]
Halostachys caspica	miR6194 and miR5655	Flavanone 3-hydroxylase	Flavonols, anthocyanidins proanthocyanidins synthesis	[32]
Podophyllum hexandrum	miR1873/miR5532	Dihydroflavonol 4-reductase C/-hydroxyisoflavanone dehydratase	Flavonoid/isoflavonoid biosynthesis	[33]
Alkaloids				
Opium poppy (*Papaver somniferum*)	pso-miR13, pso-miR2161, and pso-miR408	7-O-methyltransferase, *S*-adenosyl-L-methionine:3′-hydroxy-*N*-methylcoclaurine 4′-O-methyltransferase 2/ 4′-O-methyltransferase2 (4-OMT)/FAD-binding and BBE domain-containing protein, also known as reticuline oxidase-like protein	Benzylisoquinoline alkaloids	[34]
Tobacco	miRX17, miRX27, miRX20, and miRX19	QPT1, QPT2, CYP82E4, and PMT2	Nicotine biosynthesis and catabolism	[35]
T. baccata	miR164 and miR171	Taxane 13α hydroxylase and taxane 2α-O-benzcyltransferase	Paclitaxel biosynthetic genes	[36]
R. serpentina	miR396b	Targets kaempferol 3-O-beta-D-galactosyltransferase	Flavonol glycosides	[37]
Mentha spp.	miR156	Basic helix-loop-helix (bHLH)	Flavone/flavonol biosynthesis	[38]
Terpenoids				
P. kurroa	iRNA-4995	3-Deoxy-7-phosphoheptulonate synthase (DAHP synthase)	Terpenoid biosynthesis ultimately affecting the production of picroside-I	[39]

Plant species	miRNA	Target	Function	References
Korean ginseng (*Panax ginseng* Meyer)	miR854b and miR854c	3-Hydroxy-3-methylglutaryl-coenzyme A reductase (HMGR), farnesyl diphosphate synthase (FPS), geranyl-diphosphate synthase, squalene synthase, and squalene epoxidase (SE)		[40]
C. roseus	mir-5021	MYB transcription factor, geranyl diphosphate synthase, GCPE protein, UDP-glucose iridoid glucosyltransferase	Primary and secondary metabolism, Isoprenoid/terpenoid biosynthesis iridoid production in higher plants	[41]
Xanthium strumarium L.	miR7539, miR5021, and miR1134	Nontranscriptional factor proteins, such as DXS, HMGR, IDS, and IDI, essential to produce IPP and DMAPP	Terpenoid biosynthesis	[42]
Xanthium strumarium L.	miR7540, miR5183, miR6449, miR5255, miR5491, and miR6435	R-linalool synthase, gibberellin 3-oxidase, ent-kaurene synthase, squalene epoxidase, beta-amyrin synthase, and germacrene A oxidase	Mono-, sesqui-, di-, and tri-terpenoids biosynthesis	[42]
Ferula gummosa	miR2919, miR5251, miR838, miR5021, and miR5658	SPL7, SPL11, and ATHB13 TFs	Terpene biosynthesis	[43]
Pogostemon cablin	miRNA156	Squamosa promoter binding protein-like (SPL)	Sesquiterpene biosynthesis	[44]
A. thaliana	miR156	SPL transcription factor	Modulate sesquiterpene synthase gene TPS21 responsible for the biosynthesis of (E)-β-caryophyllene	[44]
Mentha spp.	miR156, miR414, and miR5021	Basic helix-loop-helix (bHLH) geranyl di-phosphate synthase subunit alpha-like protein (NACA), respectively	Terpenoid backbone biosynthesis, sesquiterpenoid and triterpenoid biosynthesis	[38]
Others				
S. chirayita	miR-168, miR-11320, miR-166a, miR-11071, miR-156a and miR-166b	Acetyl-CoA acetyltransferase (AACT), aspartate aminotransferase (PHAT), premnaspirodiene oxygenase (PSO), ribulose-phosphate 3-epimerase (RPE), phosphoglycerate mutase (PGM), and a gene encoding homeobox-leucine zipper protein (HD-ZIF)	Secondary metabolites biosynthesis	[23]
A. thaliana	miR163	Family of small molecules of methyltransferases	Secondary metabolism	[45]
A. thaliana	miR393	Auxin receptors (TIR1, AFB2 and AFB3)	Increase of glucosinolate and decrease of camalexin	[46]

Plant species	miRNA	Target	Function	References
Potato	mirn79	AP2/ERF transcription factor	JA-responsive secondary metabolites	[24]
S. rebaudiana	miRstv_7	UDP-glycosyltransferase 76G1 (ugt76g1), kaurenoic acid hydroxylase (KAH), and kaurene oxidase (KO)	Steviol glycoside biosynthesis	[47]
Arabidopsis thaliana	miR826 and miR5090	AOP2	Glucosinolate biosynthesis	[48]
A. thaliana	miR826	Alkenyl hydroxyalkyl producing 2	Glucosinolate synthesis	[49]
Salvia miltiorrhiza	miR5072	Acetyl-CoA C-acetyl transferase	Tanshinones biosynthesis	[50]

Table 1.
miRNA related to secondary metabolite production.

destabilization of a MYB-bHLH-WD40 transcriptional activation complex. *Diospyros kaki* fruits collected at two examined stages (15 and 20 WAF) showed differential expression of the mRNAs, indicating that these miRNAs might regulate PA synthesis during development, and some of them are miR858 and miR156, which regulate PA synthesis. miR858 positively regulates the genes responsible for the production of PA, while miR156 does so in a negative way. miR395 is another miRNA that has an influence on PA biosynthesis [30]. Some miRNAs (U436803, U977315, U805963, U3938865, and U4351355) regulate fatty acid and flavonoid biosynthesis in *Lonicera japonica* [2, 3]. The characterization in *A. thaliana* shows that miR858a targets MYB transcription factors that are involved in flavonoid biosynthesis, growth, and development. Over-expression of miR858a downregulates several MYB transcription factors, and the higher expression of MYBs in MIM858 lines leads to the redirection of the metabolic flux toward the synthesis of flavonoids [31]. Yang et al. [32] indicate that salt stress conditions regulate miRNAs; some salt stress-related biological pathways includes calcium signaling pathway, MAPK signaling pathway, plant hormone signal transduction, and flavonoid biosynthesis [32]. Himalayan mayapple (*Podophyllum hexandrum*), miR1438 target caffeoyl-CoA O-methyltransferase and is related to phenylalanine metabolism, phenylpropanoid biosynthesis, flavonoid biosynthesis, stilbenoid, diarylheptanoid, and gingerol biosynthesis. miR1873 targets dihydroflavonol 4-reductase C related to flavonoid biosynthesis. miR5532 2-hydroxyisoflavanone dehydratase is related to isoflavonoid biosynthesis [33].

2.2 Alkaloids

Alkaloids are naturally compounds that have one or more of their nitrogen atoms. Alkaloids are classified into different groups: indole, piperidine, tropane, purine, pyrrolizidine, imidazole, quinolizidine, isoquinoline, and pyrrolidine alkaloids [52]. Because of their toxicity, alkaloids act as defense compounds against diverse pathogens or herbivores. Understanding the regulation of alkaloid biosynthesis is crucial for its production. Target transcript identification analyses in Opium poppy (*Papaver somniferum*) revealed that pso-miR13, pso-miR2161, and pso-miR408 (**Table 1**) might be involved in BIA biosynthesis. pso-miR13 might cleave 7-OMT transcript, involved in the conversion of S-reticuline to morphinan alkaloids. 4-OMT is the target of pso-miR216 and mediates the production of S-reticuline that is also an intermediate molecule in BIA biosynthesis. On the other hand, pso-miR408 possibly targets mRNA from reticuline oxidase-like protein in charge of the conversion of S-reticuline to (S)-scoulerine in the BIA pathway [34]. Studies in tobacco (*Nicotiana tabacum*), identified four unique tobacco-specific miRNAs miRX17, miRX27, miRX20, and miRX19 that were predicted to target key genes of the nicotine biosynthesis and catabolism pathways, QPT1, QPT2, CYP82E4, and PMT2 genes, respectively [35]. In *Taxus baccata*, two paclitaxel biosynthetic genes, taxane 13α hydroxylase and taxane 2α-O-benzoyltransferase, are the cleavage targets of miR164 and miR171, respectively [36]. *In silico* analysis reveals that miR396b in *Rauwolfia serpentina* targets kaempferol 3-O-beta-D-galactosyltransferase whose activity as transferase activity, transferring hexosyl groups is essential for formation of flavonol glycosides [37]. A computational approach in *Mentha* spp., revealed that miR156, miR414, and miR5021 are essential for regulation of essential oil biosynthesis. miR156 participates in flavone, flavonol biosynthesis, and terpenoid backbone biosynthesis [38].

2.3 Terpenoids

Plant terpenoids secondary metabolites are synthesized from C5 precursors isopentenyl diphosphate (IPP) and dimethylallyl diphosphate (DMAPP). They are classified according to the number of carbon atoms as monoterpenes (C10), sesquiterpenes (C15), diterpenes (C20), triterpenes (C30), carotenoids (C40), and polyprenols (>45) [53]. Like the alkaloids and the flavonoids, the biological characteristics and the applications of interest in the industry. Computational identification of miRNAs was done in six transcriptomes of *Picrorhiza kurroa* revealed that miRNA-4995 has a regulatory role in terpenoid biosynthesis (**Table 1**), ultimately affecting the production of picroside-I [39]. *In silico* profiling of microRNAs (miRNAs) in Korean ginseng (*Panax ginseng* Meyer) indicate that 3-hydroxy-3-methylglutaryl-coenzyme A reductase (HMGR), farnesyl diphosphate synthase (FPS), geranyl-diphosphate synthase, squalene synthase, and squalene epoxidase (SE) were predicted been regulated by miR854b and miR854c, especially SE [40]. miR-5021 was identified in *Catharanthus roseus* which targets two enzymes involved in biosynthesis of terpenoid indole alkaloids (TIAs), GCPE protein, and Terpenoid cyclase [41]. miR7539, miR5021, and miR1134 might be involved in regulating terpenoid biosynthesis by targeting upstream terpenoid pathway genes; nontranscriptional factor proteins, such as DXS, HMGR, IDS, and IDI, essential to produce IPP and DMAPP, the common precursors for all the downstream end terpenoids [42]. miRNAs miR7540, miR5183, miR6449, miR5255, miR5491, and miR6435 target downstream enzymes in the biosynthesis of mono-, sesqui-, di-, and tri-terpenoids; they were R-linalool synthase, gibberellin 3-oxidase, ent-kaurene synthase, squalene epoxidase, beta-amyrin synthase, and germacrene A oxidase [42]. miR2919, miR5251, miR838, miR5021, and miR5658 were found to be related to the pathway of terpene biosynthesis in *Ferula gummosa*. SPL7, SPL11, and ATHB13 TFs are putatively regulated by miR1533, miR5021, and miR5658, respectively [43]. miRNA156-targeted squamosa promoter binding protein-like (SPL) intervenes in the temporal space regulation of sesquiterpene biosynthesis [44]. miR5021 is also involved in terpenoid backbone biosynthesis and miR414 is related to sesquiterpenoid and triterpenoid biosynthesis [38].

2.4 Other secondary metabolites

miRNAs were identified from *in vitro* culture of roots and leaves tissues of the transcriptome of *Withania somnifera*; miR159, miR172, miR5140, and miR5303 in root tissue and miR477, miR530, miR1426, and miR5079 of leaf tissue. These miRNA were associated in the regulation of secondary metabolites. Endogenous miRNAs (miR159 and miR5140 from roots, miR477 and miR530 from leaves) may be help to increase the metabolites (withanoides) yield. Also, miR159, miR172 from roots, and miR530 from leaves were involved in the regulation of secondary metabolite associated with mRNAs [54]. *Chlorophytum borivilianum*, *Oryza sativa*, and *Arabidopsis thaliana* target gene prediction indicate that miR9662, miR894, miR172, and miR166 might be involved in regulating saponin biosynthetic pathway [55]. miR8154 and miR5298b increase taxol, phenylpropanoid, and flavonoid biosynthesis in subcultured Taxus cells [56]. *In silico* analysis indicate that miRstv_7* target ugt76g1, KAH, KO, for steviol glycoside biosynthesis [47]. In *Arabidopsis thaliana*, miR826 and miR5090 share the target AOP2, which encodes a 2-oxoglutarate-dependent dioxygenase that is involved in glucosinolate biosynthesis [48]. *Salvia miltiorrhiza* miR5072 targets acetyl-CoA C-acetyl transferase that is involved in the biosynthesis of tanshinones [50]. miR826 targets alkenyl hydroxyalkyl producing 2 oxoglutarate dioxygenase, which is involved in glucosinolate synthesis [49].

3. Conclusion

miRNAs are small molecules associated with developmental processes controlling gene expression. The mechanisms involved posttranscriptional and transductional processes. The miRNA secondary metabolism control is a relative new field of study; the knowledge of the regulation of secondary metabolism in plants will help to understand the production of these products in controlled systems. Some of these products have an important economical value because of their use in agricultural, food, and cosmetic industries making these areas (miRNA regulation) very attractive.

Acknowledgements

The authors thank Universidad De la Salle Bajío campus Campestre for financial support. MVH thank to CONACyT for postdoctoral fellowship (288773).

Author details

Vargas-Hernández Marcela[1], Vázquez-Marrufo Gerardo[1], Aguilar-Ruiz Carlos Agustín[2], González-Márquez Marco Antonio[2], Rocha Oscar[2], Cerna-Pantoja Diego[2] and Andrés Cruz-Hernández[2]*

1 Facultad de Medicina Veterinaria y Zootecnia, Centro Multidisciplinario de Estudios en Biotecnología, Universidad Michoacana de San Nicolás de Hidalgo, Michoacán, Mexico

2 Universidad De La Salle Bajío, Campus Campestre, León, Guanajuato, Mexico

*Address all correspondence to: acruz@delasalle.edu.mx; andrex1998@hotmail.com

References

[1] Treiber T, Treiber N, Meister G. Regulation of microRNA biogenesis and its crosstalk with other cellular pathways. Nature Reviews. Molecular Cell Biology. 2018;**19**(12):808. DOI: 10.1038/ s41580-018-0059-1

[2] Liu SR, Zhou JJ, Hu CG, Wei CL, Zhang JZ. MicroRNA-mediated gene silencing in plant defense and viral counter-defense. Frontiers in Microbiology. 2017;**8**:1801. DOI: 10.3389/fmicb.2017.01801

[3] Liu J, Yuan Y, Wang Y, Jiang C, Chen T, Zhu F, et al. Regulation of fatty acid and flavonoid biosynthesis by miRNAs in *Lonicera japonica*. RSC Advances. 2017;**7**(56):35426-35437. DOI: 10.1039/C7RA05800D

[4] Melo CA, Melo SA. MicroRNA biogenesis: Dicing assay. Methods in Molecular Biology. 2014;**1182**:219-226. DOI: 10.1007/978-1-4939-1062-5_20

[5] Millar AA, Waterhouse P. Plant and animal microRNAs: Similarities and differences. Functional and Integrative Genomics. 2005;**5**:129-135. DOI: 10.1007/s10142-005-0145-2

[6] Bruscella P, Bottini S, Baudesson C, Pawlotsky JM, Feray C, Trabucchi M. Viruses and miRNAs: More friends than foes. Frontiers in Microbiology. 2017;**8**:824. DOI: 10.3389/fmicb.2017.00824

[7] França GS, Hinske LC, Galante PA, Vibranovski MD. Unveiling the impact of the genomic architecture on the evolution of vertebrate microRNAs. Frontiers in Genetics. 2017;**21**(8):34. DOI: 10.3389/fgene.2017.00034. eCollection 2017

[8] Achkar NP, Cambiagno DA, Manavella PA. miRNA biogenesis: A dynamic pathway. Trends in Plant Science. 2016;**21**(12):1034-1044. DOI: 10.1016/j.tplants.2016.09.003. Epub 2016 Oct 25

[9] Eamens A, Wang MB, Smith NA, Waterhouse PM. RNA silencing in plants: Yesterday, today, and tomorrow. Plant Physiology. 2008;**147**(2):456-468. DOI: 10.1104/pp.108.117275

[10] O'Brien J, Hayder H, Zayed Y, Peng C. Overview of microRNA biogenesis, mechanisms of actions, and circulation. Frontiers in Endocrinology. 2018;**9**:402. DOI: 10.3389/fendo.2018.00402

[11] Wang L, Song X, Gu L, Li X, Cao S, Chu C, et al. NOT2 proteins promote polymerase II-dependent transcription and interact with multiple microRNA biogenesis factors in *Arabidopsis*. The Plant Cell. 2013;**25**(2):715-727. DOI: 10.1105/tpc.112.105882. Epub 2013 Feb 19

[12] Zhang S, Xie M, Ren G, Yu B. CDC5, a DNA binding protein, positively regulates posttranscriptional processing and/or transcription of primary microRNA transcripts. Proceedings of the National Academy of Sciences of the United States of America. 2013;**110**(43):17588-17593. DOI: 10.1073/pnas.1310644110. Epub 2013 Oct 7

[13] Jagadeeswaran G, Zheng Y, Sumathipala N, Jiang H, Arrese EL, Soulages JL, et al. Deep sequencing of small RNA libraries reveals dynamic regulation of conserved and novel microRNAs and microRNA-stars during silkworm development. BMC Genomics. 2010;**11**:52. DOI: 10.1186/1471-2164-11-52

[14] Catalanotto C, Cogoni C, Zardo G. MicroRNA in control of gene expression: An overview of nuclear functions. International Journal of Molecular Sciences. 2016;**17**(10):E1712. DOI: 10.3390/ijms17101712

[15] Iwakawa HO, Tomari Y. The functions of microRNAs: mRNA decay and translational repression. Trends in Cell Biology. 2015;**25**(11):651-665. DOI: 10.1016/j.tcb.2015.07.011. Epub 2015 Oct 1

[16] Singh RK, Gase K, Baldwin IT, Pandey SP. Molecular evolution and diversification of the Argonaute family of proteins in plants. BMC Plant Biology. 2015;**15**:23. DOI: 10.1186/s12870-014-0364-6

[17] Arribas-Hernández L, Marchais A, Poulsen C, Haase B, Hauptmann J, Benes V, et al. The slicer activity of ARGONAUTE1 is required specifically for the phasing, not production, of trans-acting short interfering RNAs in *Arabidopsis*. The Plant Cell. 2016;**28**:1563-1580. DOI: 10.1105/tpc.16.00121

[18] Djami-Tchatchou AT, Sanan-Mishra N, Ntushelo K, Dubery IA. Functional roles of microRNAs in agronomically important plants-potential as targets for crop improvement and protection. Frontiers in Plant Science. 2017;**8**:378. DOI: 10.3389/fpls.2017.00378. eCollection 2017

[19] Mahajan V, Mahajan A, Pagoch SS, Bedi YS, Gandhi SG. MicroRNA mediated regulation of plant secondary metabolism: An *in silico* analysis. Journal of Natural Science, Biology and Medicine. 2011;**2**(Suppl S1): 44-45

[20] Gupta OP, Karkute SG, Banerjee S, Meena NL, Dahuja A. Contemporary understanding of miRNA-based regulation of secondary metabolites biosynthesis in plants. Frontiers in Plant Science. 2017;**8**:374. DOI: 10.3389/fpls.2017.00374

[21] Tiwari R, Rana C. Plant secondary metabolites: A review. International Journal of Engineering Research and General Science. 2015;**3**:661-670

[22] Sabzehzari M, Naghavi M. Phyto-miRNAs-based regulation of metabolites biosynthesis in medicinal plants. Gene. 2019;**682**:13-24. DOI: 10.1016/j.gene.2018.09.049. Epub 2018 Sep 26

[23] Padhan JK, Kumar P, Sood H, Chauhan RS. Prospecting NGS transcriptomes to assess regulation of miRNA-mediated secondary metabolites biosynthesis in *Swertia chirayita*, a medicinal herb of North-Western Himalayas. Medicinal Plants—International Journal of Phytomedicines and Related Industries. 2016;**8**:219-228. DOI: 10.5958/0975-6892.2016.00029.0

[24] Qiao Y, Zhang J, Zhang J, Wang Z, Ran A, Guo H, et al. Integrated RNA-seq and sRNA-seq analysis reveals miRNA effects on secondary metabolism in *Solanum tuberosum* L. Molecular Genetics and Genomics. 2017;**292**:37-52. DOI: 10.1007/s00438-016-1253-5. Epub 2016 Sep 27

[25] Jian H, Yang B, Zhang A, Ma J, Ding Y, Chen Z, et al. Genome-wide identification of microRNAs in response to cadmium stress in oilseed rape (*Brassica napus* L.) using high-throughput sequencing. International Journal of Molecular Sciences. 2018;**19**:1431. DOI: 10.3390/ijms19051431

[26] Bazzini A, Manacorda C, Tohge T, Conti G, Rodriguez MC, Nunes-Nesi A, et al. Metabolic and miRNA profiling of TMV infected plants reveals biphasic temporal changes. PLoS One. 2011;**6**:e28466. DOI: 10.1371/journal.pone.0028466

[27] Kumar S, Pandey AK. Chemistry and biological activities of flavonoids: An overview. Scientific World Journal. 2013;**2013**:162750. DOI: 10.1155/2013/162750. eCollection 2013

[28] Panche AN, Diwan AD, Chandra SR. Flavonoids: An overview. Journal

of Nutritional Science. 2016;**5**:e47. DOI: 10.1017/jns.2016.41. eCollection 2016

[29] Barozai MY, Baloch IA, Din M. Identification of microRNAs and their targets in *Helianthus*. Molecular Biology Reports. 2012;**39**:2523-2532. DOI: 10.1007/s11033-011-1004-y. Epub 2011 Jun 14

[30] Luo Y, Zhang X, Luo Z, Zhang Q, Liu J. Identification and characterization of microRNAs from Chinese pollination constant nonastringent persimmon using high-throughput sequencing. BMC Plant Biology. 2015;**15**:11. DOI: 10.1186/s12870-014-0400-6

[31] Sharma D, Tiwari M, Pandey A, Bhatia C, Sharma A, Trivedi PK. MicroRNA858 is a potential regulator of phenylpropanoid pathway and plant development. Plant Physiology. 2016;**171**:944-959. DOI: 10.1104/pp.15.01831

[32] Yang R, Zeng Y, Yi X, Zhao L, Zhang Y. Small RNA deep sequencing reveals the important role of microRNAs in the halophyte *Halostachys caspica*. Plant Biotechnology Journal. 2015;**13**:395-408. DOI: 10.1111/pbi. 12337

[33] Biswas S, Hazra S, Chattopadhyay S. Plant Gene. 2016;**6**:82-89. DOI: 10.1016/j.plgene.2016. 04.002Identification of conserved miRNAs and their putative target genes in *Podophyllum hexandrum* (Himalayan Mayapple)

[34] Boke H, Ozhuner E, Turktas M, Parmaksiz I, Ozcan S, Unver T. Regulation of the alkaloid biosynthesis by miRNA in opium poppy. Plant Biotechnology Journal. 2015;**13**:409-420. DOI: 10.1111/pbi.12346

[35] Li F, Wang W, Zhao N, Xiao B, Cao P, Wu X, et al. Regulation of nicotine biosynthesis by endogenous target mimicry of microRNA in tobacco. Plant Physiology. 2015;**169**:1062-1071. DOI: 10.1104/pp.15.00649

[36] Hao DC, Yang L, Xiao PG, Liu M. Identification of Taxus microRNAs and their targets with high-throughput sequencing and degradome analysis. Physiologia Plantarum. 2012;**146**:388-403. DOI: 10.1111/j.1399-3054.2012.01668.x. Epub 2012 Jul 25

[37] Prakash P, Rajakani R, Gupta V. Transcriptome-wide identification of *Rauvolfia serpentina* microRNAs and prediction of their potential targets. Computational Biology and Chemistry. 2016;**61**:62-74. DOI: 10.1016/j.compbiolchem.2015. 12.002

[38] Singh N, Srivastava S, Shasany A, Sharma A. Identification of miRNAs and their targets involved in the secondary metabolic pathways of *Mentha* spp. Computational Biology and Chemistry. 2016;**64**:154-162. DOI: 10.1016/j.compbiolchem.2016.06.004

[39] Vashisht I, Mishra P, Pal T, Chanumolu S, Singh T, Chauhan R. Mining NGS transcriptomes for miRNAs and dissecting their role in regulating growth, development, and secondary metabolites production in different organs of a medicinal herb, *Picrorhiza kurroa*. Planta. 2015;**241**: 1255-1268. DOI: 10.1007/s00425-015-2255-y. Epub 2015 Feb 7

[40] Mathiyalagan R, Subramaniyam S, Natarajan S, Kim YJ, Sun MS, Kim SY, et al. *In silico* profiling of microRNAs in Korean ginseng (*Panax ginseng* Meyer). Journal of Ginseng Research. 2013;**37**:227-247. DOI: 10.5142/jgr.2013.37.227

[41] Pani A, Mahapatra RK. Computational identification of microRNAs and their targets in *Catharanthus roseus* expressed sequence tags. Genome Data. 2013;**1**:2-6. DOI: 10.1016/j.gdata.2013.06.001

[42] Fan R, Li Y, Li C, Zhang Y. Differential microRNA analysis of glandular trichomes and young leaves in *Xanthium strumarium* L. reveals their putative roles in regulating terpenoid biosynthesis. PLoS One. 2015;**25**(10):e0139002. DOI: 10.1371/journal.pone.0139002. eCollection

[43] Sobhani NA, Naghavi MR. Mining *Ferula gummosa* transcriptome to identify miRNAs involved in the regulation and biosynthesis of terpenes. Gene. 2018;**1**(645):41-47. DOI: 10.1016/j.gene.2017.12.035. Epub 2017 Dec 19

[44] Yu Z, Wang L, Zhao B, Shan C, Zhang Y, Chen D, et al. Progressive regulation of sesquiterpene biosynthesis in *Arabidopsis* and *Patchouli* (*Pogostemon cablin*) by the miR156-targeted SPL transcription factors. International Journal of Engineering Research and General Science. 2015;**8**:98-110. DOI: 10.1016/j.molp.2014.11.002

[45] Ng D, Zhang C, Miller M, Palmer G, Whiteley M, Tholl D, et al. Cis- and trans-regulation of miR163 and target genes confers natural variation of secondary metabolites in two *Arabidopsis* species and their allopolyploids. The Plant Cell. 2011;**23**:1729-1740. DOI: 10.1105/tpc.111.083915. Epub 2011 May 20

[46] Seilaniantz A, MacLean D, Jikumaru Y, Hill L, Yamaguchi S, Kamiya Y, et al. The microRNA miR393 re-directs secondary metabolite biosynthesis away from camalexin and towards glucosinolates. The Plant Journal. 2011;**67**:218-231. DOI: 10.1111/j.1365-313X.2011.04591.x. Epub 2011 Jun 24

[47] Saifi M, Nasrullah N, Ahmad M, Ali A, Khan J, Abdin M. *In silico* analysis and expression profiling of miRNAs targeting genes of steviol glycosides biosynthetic pathway and their relationship with steviol glycosides content in different tissues of *Stevia rebaudiana*. Plant Physiology and Biochemistry. 2015;**94**:57-64. DOI: 10.1016/j.plaphy.2015.05.009

[48] He H, Liang G, Li Y, Wang F, Yu D. Two young microRNAs originating from target duplication mediate nitrogen starvation adaptation via regulation of glucosinolate synthesis in *Arabidopsis thaliana*. Plant Physiology. 2014;**164**(2):853-865. DOI: 10.1104/pp.113.228635. Epub 2013 Dec 23

[49] Liang G, He H, Yu D. Identification of nitrogen starvation-responsive microRNAs in *Arabidopsis thaliana*. PLoS One. 2012;**7**:e48951. DOI: 10.1371/journal.pone.0048951. Epub 2012 Nov 14

[50] Xu X, Jiang Q, Ma X, Ying Q, Shen B, Qian Y, et al. Deep sequencing identifies tissue-specific microRNAs and their target genes involving in the biosynthesis of tanshinones in *Salvia miltiorrhiza*. PLoS One. 2014;**9**:e111679. DOI: 10.1371/journal.pone.0111679

[51] Gou JY, Felippes FF, Liu CJ, Weigel D, Wang JW. Negative regulation of anthocyanin biosynthesis in *Arabidopsis* by a miR156-targeted SPL transcription factor. The Plant Cell. 2011;**23**:1512-1522. DOI: 10.1105/tpc.111.084525

[52] Roy A. A review on the alkaloids an important therapeutic compound from plants. Plant Biotechnology Journal. 2017;**3**:1-9. DOI: 10.1021/acs.jpcb.5b10121

[53] Jiang Z, Kempinski C, Chappell J. Extraction and analysis of terpenes/terpenoids. Current Protocols in Plant Biology. 2016;**1**:345-358. DOI: 10.1002/cppb.20024

[54] Swati-Srivastava, Sanchita, Singh R, Srivastava G, Sharma A. Comparative study of withanolide biosynthesis-related miRNAs in root and leaf tissues of *Withania somnifera*. Applied Biochemistry and Biotechnology. 2018;**185**:1145-1159. DOI: 10.1007/s12010-018-2702-x. Epub 2018 Feb 24

[55] Kajal M, Singh K. Small RNA profiling for identification of miRNAs involved in regulation of saponins biosynthesis in *Chlorophytum borivilianum*. BMC Plant Biology. 2017;**17**:265. DOI: 10.1186/s12870-017-1214-0

[56] Zhang M, Dong Y, Nie L, Lu M, Fu C, Yu L. High-throughput sequencing reveals miRNA effects on the primary and secondary production properties in long-term subcultured Taxus cells. Frontiers in Plant Science. 2015;**6**:604. DOI: 10.3389/fpls.2015.00604

Section 3

Biological Activity of Phenolic Compounds

Chapter 5

Bioavailability and Metabolic Pathway of Phenolic Compounds

Muhammad Bilal Hussain, Sadia Hassan, Marwa Waheed, Ahsan Javed, Muhammad Adil Farooq and Ali Tahir

Abstract

As potential agents for preventing different oxidative stress-related diseases, phenolic compounds have attracted increasing attention with the passage of time. Intake of fruits, vegetables and cereals in higher quantities is linked with decreased chances of chronic diseases. In plant-based foods, phenolic compounds are very abundant. However, bio-accessibility and biotransformation of phenolic compound are not reviewed in these studies; therefore, a detailed action mechanism of phenolic compounds is not recognized. In this article, inclusive concept of different factors affecting the bioavailability of phenolic compounds and their metabolic processes is presented through which phenolic compounds go after ingestion.

Keywords: polyphenols, bioavailability, biotransformation, metabolism

1. Introduction

In recent past, the awareness of the consumer related to the effect of diet on the health has been improved; therefore, leading to upsurge in the consumption of vegetables, cereal based foods and fruits. Numerous studies have suggested the bioactive characteristics of the bioactive moieties, i.e., phenolic compounds. Nonetheless, bioactive claims are made without taking into consideration the further modifications to which phenolic compounds are subjected once ingested [1].

Phenolic compounds are the secondary metabolites of plants which constitute an important group, i.e., phenylpropanoids. These compounds possess an aromatic ring and various OH groups which are link to it. On the basis of classification, phenolic compounds are prorated into various subgroups. They are grouped as a function of the number of phenolic rings that they contain and the radicals that bind these rings to another one [2, 3]. Phenolic compounds have fetched substantial focus as the ingestion of these bioactive moieties is correlated to lower the prevalence of chronic ailments, for example, diabetes, CVD and cancer. Cereals, fruits, and vegetables are rich sources of phenolic compounds. In fact, the health benefits of their dietary intake have been related, at least in part, to their phenolic compounds content [4]. This manuscript presents the bird's eye view of the health claims as well as bioavailability of the phenolic compounds.

 IntechOpen

2. Dietary phenolic compounds

Phenolic compounds are the derivatives of secondary metabolism of plants. Chemically phenolic compounds consist of aromatic ring to which one or more OH^- substituents are attached [1, 5]. Despite of diversity of phenolic compounds, they are mainly divided into two subgroups, (1) flavonoids and (2) non-flavonoids. First one constitutes of heterocyclic oxygen which are bonded with two aromatic rings and depends on the amount of hydrogenation. They can be further subdivided into six subgroups, i.e., flavanol, flavones, anthocyanins, flavonols, flavanones and isoflavones. While the later one, like cinnamic and benzoic compounds, they contain aromatic ring which are attached to organic acids. Lignins, stilbenes and tannins are also the subgroups of non-flavonoid compounds. Characteristics like flavor, astringency and color are instigated due to presence of these compounds [1].

3. Food sources with reported bioactivity

Latterly, due to numerous health prompting effects, for example, antimicrobial [6], neuro-protective [7], antioxidant [8], cardioprotective [9], anti-inflammatory [10] and cancer preventive [11] properties, phenolic compounds have much of the attention of the researchers. Phenolic compounds possess different derivatives which have a potential application in the prevention or treatment of these aliments [12]. Likewise, Perez-Vizcaino et al. reported numerous studies which supports the fact that upsurge in the consumption of foods rich in the phenolic compounds might be linked with the prevention of above-mentioned disorders [13]. Vegetables, fruits and cereals have high concentration of the phenolics. **Table 1** includes examples of some foods rich in phenolic compounds with reported biological effects.

Phenolic compound	Source	References
Phenolic acids (gallic acid)	Red wine	[14]
Anthocyanins (cyanidin, delphinidin, malvidin, pelargonidin, peonidin)	Blackberry, blueberry, black grape, cherry, strawberry, red wine, plum	[15]
Condensed tannins (procyanidin)	Red wine, chocolate, cranberry juice and apples	[16]
Flavan-3-ols (catechin)	Fruits, vegetables, chocolate, lentil, green and black tea, wine, grapes and ginkgo	[17, 18]
Flavanones (hespertin, naringenin)	Orange, grapefruit and lemon juices	[19]
Flavones (apigenin, luteolin)	Parsley, celery, capsicum pepper and grape	[18]
Flavonols (quercetin, kaempferol)	Fruits, vegetables, and beverages such as tea and red wine	[20, 21]
Isoflavones (genistein)	Soy	[22]
Stilbenes (resveratrol)	Legumes, grapes, red wine, soy, peanuts and peanut products	[23, 24]

Table 1.
Phenolic compounds from food sources with reported biological effects.

4. Bioactivities of polyphenols

Since from the ancient times, bioactive moieties extracted from the natural sources have guarantee medicinal characteristics in combating against certain disorders [25]. These bioactive components have a wide range of potential applications, i.e., antimicrobial, antitumoral, antimicrobial, hepato-protection and antioxidant. Few phenolic compounds show higher free radical scavenging properties individually, while numerous others showed these characteristics in synergism [26]. The free radical scavenging properties generally influenced by the chemical structure, position and number of OH^- group as well as glycosylation or other forms of replacement [27]. Additionally, despite of nullification of the ROS, nitrosative and oxidative stress such as CVD, neurological disorders, diabetes mellitus, cancer, and hypertension have also been prevented by phenolic compounds [28]. Likewise, literature also showed the anti-inflammatory properties of the phenolic compounds [29]. It is observed during the inflammation several RNS (reactive nitrogen species) as well as ROS (reactive oxygen species) are formed which escalate the action of proinflammatory factors. Phenolic compounds limit the pro-inflammatory enzymes thus prevent the human body from adverse effects [30].

In both developed and developing nations, cancer is the major cause of millions of demises each year [31]. For the treatment of numerous aliments, plants are the important ally in the traditional medicine. In pharmaceutical sector natural components impart important proportion in the synthesis of new anticancer drugs [32]. In the treatment of tumor cell, the uses of synthetic moieties are linked with the toxicity problems. Carocho and Ferreira suggested that without any toxicity or side effects natural compounds extracted from plants can be administered. By using both human trial and *in vivo* models, the effects of phenolic compounds on tumor cells have been comprehensively investigated [26]. Huang et al. reported the capacity of phenolic compounds to induce apoptosis by regulate carcinogen metabolism, ontogenesis and cell cycle arrest, suppress cell adhesion and DNA binding, proliferation, migration and block signaling pathways [11]. Similarly, the effects of phenolic compounds against hepatoprotective capacity have also been comprehensively studied both *in vitro* and *in vivo*. Phenolic acid and flavonoids have fetched the attention due to high free radical scavenging properties which overcome liver injuries frequently caused due to oxidative reaction which endorse lipid peroxidation in hepatic tissues [33].

Beside above-mentioned bioactivities, polyphenols have exhibited numerous other health beneficial effects.

5. Bioavailability of phenolic compounds

It is necessary to have the knowledge about the availability of the bioactive component as they are very effect against in the prevention of the disorders. By definition, the concentration of nutrient that is ingested, absorbed and metabolized via normal passages [34]. The bioavailability profile is not directly improved by the intake of high content of phenolic compounds [35]. Rein and his fellows purposed that to guarantee the bio efficacy of phenolic compounds, bioavailability is recognized as ultimate step, for example, at dietary level; the bioavailability is the proportion of a food which is ingested and consumed and thus a matter of nutritional efficacy [36]. Hence, numerous other factors affected may impart interference in the direct bioavailability of the phenolic compounds present in the food. Examples of several external aspects are interaction with other moieties, food processing and various other intestinal factors [37]. Likewise, different and complex processes, i.e., distribution, liberation, elimination, absorption and metabolism phases also affect the bioavailability whereas, limiting factor, i.e., intestinal level absorption decreased the bioavailability [36].

Through the GIT tract, gallic acid and isoflavones which has small molecular weight are easily absorbed [35]. On the other hand, numerous phenolic compounds absorbed at a rate of 0.3–43% and the metabolite content circulating in the plasma can be low [36]. Likewise, kaempferol and quercetin belongs to flavonols exhibited several biological *in vivo* effects [38]. Yet, the utilization of these compounds as a potential health promoting components has inadequate efficiency due to the lessen bioavailability as a result of low absorption rate, low water solubility and increase instability in alkaline and neutral media including various organs, i.e., colon, small intestine, kidney and colon [39]. Due to low solubility and instability the use of apigenin is also limited in the pure form [40].

The bioavailability of phenolic compounds found in the foods begins in the oral cavity via metabolism reactions. In food transformation, mechanical action, i.e., mastication, impart significant role in the disruption of the food components which releases the compounds. In the oral cavity, the metabolism of glycosylated phenolic compounds commences immediately as the come in contact with the glycosidase enzymes of bacteria [41]. Literature showed that anthocyanin present in the fruit extract rich in phenolic compounds and human saliva, were moderately metabolized by the oral microflora enzymes [42]. During the passage through the stomach, few compounds go through the hydrolysis while on the other hand numerous polyphenols remain intact. Correa-Betanzo and his fellows reported that stability and modification of these food components were interlinked with the reaction of intestinal microbiota. In the G.I tract, various phenolic components need structural modification for their absorption [43]. Few *ex vivo* studies indicated that phenolic acid absorption took place in intestinal portion, i.e., jejunum and colon or at the gastric level [44]. Certain chemical characteristics, for example, molecular weight, lipophilicity, stereochemistry and the presence of group capable of hydrogen bonding, affect the transport and permeability of the polyphenols into the cytosol enterocytes from the gut lumen [45]. It is believed that the phenolic compounds are absorbed by a passive diffusion mechanism or by carriers present in the intestine, such as P-glycoprotein and cotransporters for SGLT1. These transporters are expressed on the cell membrane and transport the drugs into the cell interior [46, 47]. For example, aglycones cross the membrane of the epithelial cells via passive diffusion [41], whereas on the other hand glycosides, esters and polymers cannot cross the membrane by passive diffusion.

As long as biotransformation reaction as concerned, liver is recognized as the main organ in which maximum glycosylated phenolic compounds are metabolized by the action of small intestinal microbiota enzymes as well as by the intestinal cell membrane hydrolases, i.e., lactase phlorizin hydrolase. These first passage reactions took place in the intestine, letting the prior metabolism of compounds, which, in turn, encourage absorption. From the colon these compounds are then transported to the liver through portal vein or distributed in the bloodstream by the plasma proteins. During the metabolism, variation in the concentration of substance in the blood is influenced by the structural changes and absorption. For instance, plasma protein, for example, albumin transports phenolic substances along with their metabolites [48]. Meanwhile in the liver, phenolic compounds are further bio-transformed which aim to make them more polar molecules, assisting in their excretion. In the liver, these biotransformation processes are mainly categorized into two phases, i.e., phase I includes oxidation and reduction, hydrolysis reactions which are catalyzed by the CYP450 enzymes [46], while phase II enhances the hydrophilicity of the molecules prior to their elimination [41].

5.1 Factors affecting the bioavailability

Variation in the phenolic bioavailability ranges from 0.3% in the case of anthocyanins to 43% estimated for isoflavones [49]. In this sense, **Table 2** is imperative

Type of factor			References
Phenolics related factors	Chemical structure	Chemical structure solubility bond with sugars (glycosides), methyl groups, etc. stereo-configuration.	[49]
	Interaction with other compounds	Bonds with proteins (i.e., albumin) or with polyphenols with similar mechanism of absorption.	[50, 51]
Food related factors	Food processing	Thermal treatments lyophilization cooking and methods of culinary preparation storage.	[52, 53]
	Food interaction	Food matrix presence of effectors of absorption (positive or negative) (i.e., fat, fiber).	[54]
Host related factors	Dietary intake	Differences between countries and seasons quantity and frequency of exposure, single or multiple dose.	[55]
	Absorption and metabolism	Intestinal factors (i.e., enzyme activity intestinal transit time colonic microflora). Systemic factors (i.e., gender and age disorders and/or pathologies genetics physiological condition).	[55]
Other factors	Distribution and food content	Exclusivity in some foods (i.e., soy isoflavones, flavanones in citrus, etc.). Ubiquity (i.e., quercetin).	[56]
	External factors	Environmental factors (i.e., different stress conditions, degree of ripeness).	[57]

Table 2.
Factors that can affect dietary phenolic compound bioavailability.

to discuss that bioavailability is influenced by the food matrix and processing, phenolic structure and host; in addition to all these factors can interrelate with one another and effect bioavailability of the phenolic compounds, which make it tough to elaborate the particular mode of action of phenolic compounds. Additionally, pre-systemic elimination, release of a dosage form and absorption as well as route of administration also affects the bioavailability [58]. In order to elucidate the bioactive potential of phenolic compounds, *in vivo* estimation system is based on the inner wall as well as on the surface of the organs of the gastrointestinal tract, nonetheless such system did not reflect antioxidant *in vivo* effects [59], since solubility, base structure, interaction with other components as well as molecular size are the major physiochemical properties which lower the action of phenolic compounds [35].

6. Biotransformation of phenolic compounds

Transformation of lipophilic compounds into hydrophilic compounds is called biotransformation of heterologous compounds, in which the compounds are easily absorbed and excreted (**Figure 1**). The acetylation and methylation are excluded from this process, which reduces the water solubility of some alien organisms. Phenolic compounds are classified as exotic biological compounds which undertake heterologous biotransformation [61]. Heterogeneous biotransformation reactions can be divided into four categories, i.e., hydrolysis reaction, reduction reaction, oxidation reaction, and conjugation reaction. However, in study of dietary phenols, conjugation and oxidation are considered as most important reactions. Major sites of these reactions are cytoplasm, mitochondria, microsomes, and the tissues of the small intestine and liver [61]. Dietary phenolic compounds present in food have different structures as compared to those present in the tissues and peripheral circulation due to enduring metabolism after ingestion [62]. Phenol-sulfonate transferase, beta-glucosidase, lactase root enzymes hydrolase, and UDP-glucuronyl transferase are the enzymes involved in these processes. Absorption and bioavailability of phenolic

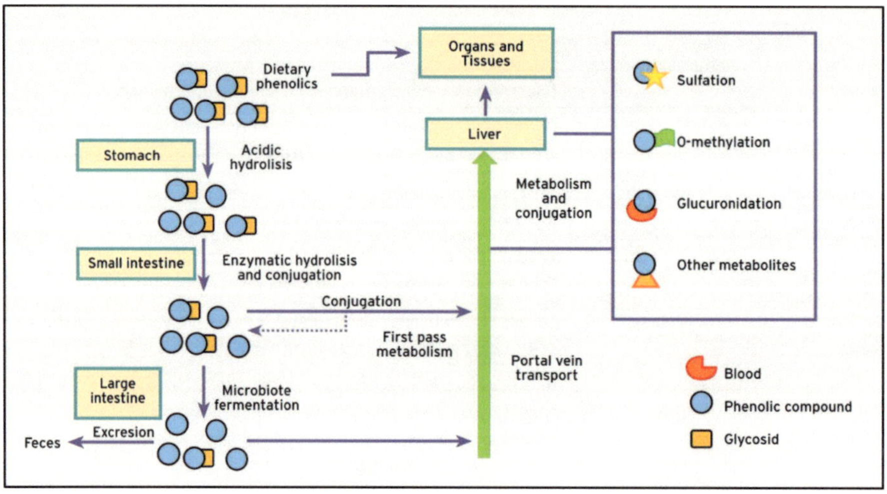

Figure 1.
Schematic representation of the metabolic pathway of phenols in humans after ingestion. Adapted from Grijalva et al. [60].

compounds mostly depend on their metabolic reactions conducted in small intestine. In addition, those compounds which are not absorbed through stomach and small intestine are degraded in large intestine through colonic microorganisms [63].

However, during chewing and digestion inn stomach, the interactions structural characters of phenolic compounds can be modified with other food substrates. This can decrease or increase small intestine's biological accessibility [41]. These binding reactions produce metabolites that bind to albumin mainly through the blood and distributed to different tissues and organs. Phenolic compounds are metabolized by colonic bacteria in parts of the intestine that are not absorbed and reach the large intestine. Colonic bacteria destroy aromatic rings and release aglycones which can be absorbed when processed into derivatives of benzoic acid and combined with glucuronic acid, glycine and sulfate [64]. Conjugated hydroxycinnamic acid *in vivo* has an effect on their biological activity because antioxidant potential of hydroxycinnamic acid is determined through existence of free hydroxyl groups, which are the important spots of sulfation and glucuronidation process [65].

7. Metabolism of phenolic compounds

7.1 Phase I

Biotransformation reactions involved during phase I of metabolism are oxidation, reduction, and hydrolysis. The biological activity of phenolic compounds can be increased, decreased or counteracted through these reactions [61]. The first stage of the reaction aims to change the structure of exogenous biomolecules. This amendment is attained through introducing amino, carboxyl and hydroxyl groups, etc. The major purpose of this reaction is to enhance the polarity of heterogeneous phenolic compounds for facilitating their excretion [61].

7.1.1 Hydrolysis

Hydrolysis is mainly aimed at functional groups such as carboxylic esters, amides, and lactones. Carboxylesterases are the most important hydrolysis enzymes in mammals. However, aldehyde dehydrogenase, carbonic anhydrase, carboxypeptidase, lipase, and protease showed hydrolytic activity [61]. Lactose-phloretin hydrolase (LPH) exists in the human body. It mainly exists on one side of the small intestinal cavity. Lactose is hydrolyzed into galactose and glucose by LPH. It is also stated that flavonoid-*O*-beta-D-glycoside can be hydrolyzed through lactose-phloretin hydrolase resulting decrease of the polarity of the aglycones produced, thereby increasing the cell absorption of flavonoids [46]. However, the function of LPH can be disabled through stearic acid factor, and it cannot further hydrolyze glycosides (e.g., rhamnoside) [63].

7.1.2 Oxidation

During phase I of biotransformation, oxidation of phenolic compounds is the process having major importance. This reaction is primarily facilitated through enzyme-based oxidation process, controlled by CYP450 (microsomal cytochrome). CYP450 in humans has an extensive variety of substrates CYP3A4 (subfamily of CYP450) in gut is responsible for metabolic reactions of exogenous. According to different reports CYP3A4 interrelates with phenolic compounds. Therefore, it is suggested that some toxic consequences can be stimulated by the combined administration of both drugs and phenolic compounds [66]. Main factor affecting the metabolic activity of CYP450 regarding phenolic compounds include binding with different phenolic compounds, functional groups, glycosylation, molecular weight, polymerization and stereo-structure [5]. In addition, hydroxyl-rich flavonoids are unlikely to be processed through CYP450; unexpectedly, catechins from tea (hydroxyl-rich flavonoids) have been reported to inhibit CYP450 [67]. Still metabolic process of phenolic compounds is under research. Yet, the possible health potential of phenols has provoked further research in this area.

7.1.3 Reduction

Majority of the absorption of polyphenol took place in the large intestine where colonic microflora imparts significant role in the catabolism of these compounds. In humans, fission of the Cring both in naringenin and quercetin were done by the enzymes produced by the *C. orbiscindens* and *E. ramulu,* whereas *E. casseliflavus* were characterized for deglycosylation of Quercetin-1-glucode [68]. With the fission of Cring, epicatechin degradation initiated which lead to the synthesis of 1-(3′,4′)-dihydroxyphenyl)-3-(2″,4″,6″-trihydroxy) propan-2-ol, which in turn transformed into 5-(3′,4′)-dihydroxyphenyl-valerolactone. Likewise, in the next step formation of 5-(3′,4′)-dihydroxyphenyl-valeric acid is done by the breakdown of valerolactone ring which further undergo beta oxidation to 3-hydroxyphenylpropionic acid whereas if alpha oxidation is done, the resultant yield is 3-hydroxyphenylacetic acid. Similarly, in epicatechin gallate and epigallocatechin gallate breakdown, the galloyl moiety is removed by the esterase enzyme and the released gallic acid is decarboxylated to pyrogallol [69]. In the liver, metabolites which are synthesize by the colon microglora are transformed to monosulfates of 5-(3′,4′)-dihydroxyphenylvaleric acid, hydroxyphenylpropionic acid and monoglucuronides through conjugation reactions. Afterword, majority of the metabolites are then transported in the blood stream and then eliminated from the body via urine [70], however the unabsorbed metabolites are excreted by the body with feces [71].

7.2 Phase II

The second stage of biotransformation involves the incorporation of different chemical radicals to exogenous compounds. In body transported free radicals are acquired from endogenous, polar, and highly available molecules. The major goal of this phase is to enhance the polarity of exogenous compounds. Increase in polarity is conducive to urination of exogenous substances [72]. Uridine 5′-diphosphate glucuronide transferase, sulfonate transferase, and catechol *O*-methyltransferase are the related enzymes for metabolism of dietary polyphenol in second phase. The synthesized molecule is bound to sulfate, glucuronic acid and/or methylation group [61]. Phenol conjugated compounds are different from their parent molecules regarding their ionic form, polarity and size. Thus, they are also different physiologically as compared to natural compounds. As a result, there is a growing need to know the possible health potential of these compounds either through *in vitro* studies.

7.2.1 Glucuronidation

It's the major conjugation-based reaction in humans. Through using UDPGA (glucuronic acid diphosphate glucuronic acid) as a substrate, glucuronidation process binds to the exogenous compound like glucuronic acid. UDP glucose, UDP xylose, and UDP galactose can also be used as substrates for this reaction [61]. The enzyme responsible for catalyzing the process of glucuronidation is UDP-glucuronide transferase. This enzyme exists in the microsomal tissue of the skin, brain, kidney, liver, and small intestine [73]. Glucuronidation sites are nucleophilic heteroatoms rich in electrons (like O, N or S). Thus, in glucuronidation reaction mostly involved substrates comprise of functional groups, like phenols and aliphatic alcohols [74]. So, in human body metabolism glucuronidation is main binding reaction of phenolic compounds [75]. Steffen et al. reported the glucuronic acid metabolite (−)-epicatechin binding serum albumin is relatively less than its aglycone. Therefore, the absorption of intestinal cells can be enhanced by the activity of beta-glucuronidase or LPH. In addition, aglycones are more lipophilic than flavonoid glycosides, so they are more easily absorbed. In addition, it is of great significance to evaluate the glucuronidation process of phenolic acid and study its effects regarding bioavailability and biological activities [76].

7.2.2 Methylation

Epicatechin gallate and epicatechins are absorbed without hydrolysis or disruption of conjugated bonds [62]. Different research trials have revealed that about 50% of total epicatechin reaching the intestine is absorbed by metabolites (especially sulfate conjugates) that are cleared into the intestinal cavity, while the clearance of epicatechin is relatively mild. The (−)-epicatechin secreted by bile may also be absorbed and be cleared by efflux in another segment of the intestine [77]. As compared to conjugation process, methylation reaction differs because it normally reduces hydro solubility of phenolics and hides the functional groups to prevent them to be attacked by conjugating enzymes [61]. As mentioned earlier, flavonoids are mainly glucuronidation, however, methylation metabolites have also been detected, for example, *O*-methylation and glucuronidation have been found in perfusion studies of catechins or epicatechins. This reaction is supposed to be facilitated through enzyme COMT (catechol *O*-methyltransferase) [78].

The *O*-methyltransferase is a highly selective enzyme system in plants, microorganisms, and mammals. This enzyme is involved in *O*-methylation of flavonoids, which is a natural xenobiotic transformation [79]. Methylation of phenolic compounds significantly improves their transportation through biological membranes and makes them more stable regarding metabolic reactions. It also improves their efficacy in different biological activities, especially their anti-tumor potential in cell culture research. Thus, as compared to hydroxylated derivatives in cell culture,

O-methylated flavonoids have a much better anticancer potential, due to being more resistant to metabolic reactions in liver and having high absorption in intestine [80]. In addition, methylated flavonoids play an important role in protein transportation having major role central role in the body defense system against toxic substances like MDR proteins [81]. It has been recommended that stability and transportation ability through biological membranes is improved by enhancing methylation degree and decreasing the number of free hydroxyl groups which can be bind with glucuronic acid and sulfuric acid groups [82].

8. Applications in functional foods and nutraceutical formulations

Nowadays, some functional foods and health products can be found commercially to capture consumers' interests. The scientific community can prove the beneficial health effects of these products, and the food and pharmaceutical industries have followed suit to develop feasible production of novel high-yield products [83]. Medicinal and aromatic plants play an important role in the field because

Phenolic compound	Source	Application	Bioactivity		References
			Antioxidant	Antimicrobial	
Phenolic acids	Commercial	Infant cereals	✓	_	[85]
Anthocyanins	Cranberry (*Vaccinium macrocarpon* Ait.)	Nutraceutical capsules	✓		[86]
Phenolic extracts	Blackberry flower (*Rubus ulmifolius* Schott)	Yogurt	✓	_	[87]
	Borage (*Borago officinalis* L.)	Fresh pasta	_	✓	[83]
	Chamomile flower (*Matricaria recutita* L.)	Cottage cheese and yogurt	✓	✓	[88, 89]
	Fennel aerial parts (*Foeniculum vulgare* Mill.)	Cottage cheese and yogurt	✓	✓	[88, 90]
	Garcinia fruit (*Garcinia cowa* Roxb)	Bread	✓	_	[91]
	Grape seed (*Vitis vinífera* L.)	Yogurt	✓	_	[92]
	Green tea (*Cammelia sinensis* L.)	Bread	✓	_	[93]
	Guava flower (*Psidium guajava*)	Bread melanoidins	✓	_	[94]
	Pomegranate fruit (*Punica granatum* L.)	Yogurt and pasta	✓	_	[95, 96]
	Pomegranate peels (*Punica granatum* L.)	Ice cream	✓	_	[97]
	Rosemary (*Rosmarinus officinalis* L.)	Cottage cheese	✓	_	[98]
	Veronicas (*Veronica montana* L.)	Cream cheese	_	✓	[99]

Table 3.
Phenolic compounds used as nutraceuticals or bioactive compounds in functional foods.

of their antioxidant, antimicrobial, and other beneficial effects in the prevention and treatment of certain diseases. The incorporation of these compounds in food can be carried out directly in free form; however, microencapsulation technology has emerged as a very effective and promising strategy to ensure the bioavailability of these compounds and help overcome the problems of food processing and intake [84]. After ingestion, these compounds are absorbed into the blood, causing changes in various cellular mechanisms, thus preventing various diseases. Many kinds of literature have proved the biological activity of phenolic compounds in various plants and fruits, and few studies have reported its application in the development of functional food or nutritional preparations (**Table 3**).

9. Potential toxicity

In recent past years, the potential toxicity of some polyphenols, such as catechins, to DNA of mouse spleen cells has been reported. DNA can be damaged due to high concentration of catechin on spleen cells of mice [100]. In addition, grape extract could also promote sister chromatid exchange induced by mitomycin C in human peripheral blood lymphocyte at a concentration of 75–300 μg/mL [101]. At the same concentration, the mixture of caffeic acid, gallic acid, and rutin hydrate could enhance mitomycin C induced fragmentation. In addition, after 24 h or more of high concentration epicatechin treatment, there was a significant negative effect on fibroblasts and keratinocytes. In addition, compounds with gallate groups showed more potential toxicity than compounds without gallate groups [102]. The results showed that polyphenols could play a positive role in the safe concentration range. However, polyphenol concentration is not the only determinant, and its negative effects are related to synergistic effects and exposure time. Therefore, the dosage and composition of polyphenols should be further studied for safe and healthy application [103].

10. Conclusion

Phenolic compounds, group of antioxidant phytochemicals have health promoting effects and potential to decrease the chances of chronic diseases linked with high consumption of fruits, vegetables and cereals. Several epidemiological studies relate health promoting effects of plant-based foods to phenolic compounds. Still, there is a deficiency of studies and research work regarding metabolic pathway of these compounds; leading to a less understanding of their mechanism of action. Therefore, it's essential to conduct further research to create improved approaches to take advantage of health promoting effects of these compounds.

Conflicts of interest

No competing interest.

Author details

Muhammad Bilal Hussain[1], Sadia Hassan[1,2]*, Marwa Waheed[1], Ahsan Javed[1], Muhammad Adil Farooq[3] and Ali Tahir[1]

1 Institute of Home and Food Sciences, Faculty of Life Sciences, Government College University, Faisalabad, Pakistan

2 Food Processing Centre, Department of Food Science and Technology, University of Nebraska-Lincoln, Nebraska, USA

3 School of Food Science and Engineering, Guangdong Province Key Laboratory for Green Processing of Natural Products and Product Safety, South China University of Technology, Guangzhou, China

*Address all correspondence to: sadiahassan88@gmail.com

References

[1] Gutiérrez-Grijalva EP, Ambriz-Pére DL, Leyva-López N, Castillo-López RI, Heredia JB. Dietary phenolic compounds, health benefits and bioaccessibility. Archivos Latinoamericanos de Nutricion. 2016;**66**(2):87-100

[2] Manach C, Williamson G, Morand C, Scalbert A, Rémésy C. Bioavailability and bioefficacy of polyphenols in humans. I. Review of 97 bioavailability studies. The American Journal of Clinical Nutrition. 2005;**81**(1):230S-242S

[3] Vermerris W, Nicholson R. Families of phenolic compounds and means of classification. Phenolic Compound Biochemistry. In: Phenolic compound biochemistry. Dordrecht: Springer; 2008. pp. 1-34

[4] Ozcan T, Akpinar-Bayizit A, Yilmaz-Ersan L, Delikanli B. Phenolics in human health. International Journal of Chemical Engineering and Applications. 2014;**5**(5):393

[5] Bravo L. Polyphenols: Chemistry, dietary sources, metabolism, and nutritional significance. Nutrition Reviews. 1998;**56**(11):317-333

[6] Chakraborty M, Mitra A. The antioxidant and antimicrobial properties of the methanolic extract from *Cocos nucifera* mesocarp. Food Chemistry. 2008;**107**(3):994-999

[7] Tavares L, Figueira I, Macedo D, McDougall GJ, Leitão MC, Vieira HL, et al. Neuroprotective effect of blackberry (*Rubus* sp.) polyphenols is potentiated after simulated gastrointestinal digestion. Food Chemistry. 2012;**131**(4):1443-1452

[8] Xiao Z, Fang L, Niu Y, Yu H. Effect of cultivar and variety on phenolic compounds and antioxidant activity of cherry wine. Food Chemistry. 2015;**186**:69-73

[9] Quiñones M, Miguel M, Aleixandre A. Beneficial effects of polyphenols on cardiovascular disease. Pharmacological Research. 2013;**68**(1):125-131

[10] Martin DA, Taheri R, Brand MH, Draghi A II, Sylvester FA, Bolling BW. Anti-inflammatory activity of *aronia* berry extracts in murine splenocytes. Journal of Functional Foods. 2014;**8**:68-75

[11] Huang WY, Cai YZ, Zhang Y. Natural phenolic compounds from medicinal herbs and dietary plants: Potential use for cancer prevention. Nutrition and Cancer. 2009;**62**(1):1-20

[12] Kishimoto Y, Tani M, Kondo K. Pleiotropic preventive effects of dietary polyphenols in cardiovascular diseases. European Journal of Clinical Nutrition. 2013;**67**(5):532

[13] Perez-Vizcaino F, Duarte J, Jimenez R, Santos-Buelga C, Osuna A. Antihypertensive effects of the flavonoid quercetin. Pharmacological Reports. 2009;**61**(1):6775

[14] Teissedre PL, Landrault N. Wine phenolics: Contribution to dietary intake and bioavailability. Food Research International. 2000;**33**(6):461-467

[15] Clifford MN. Anthocyanins—Nature, occurrence and dietary burden. Journal of the Science of Food and Agriculture. 2000;**80**(7):1063-1072

[16] Hammerstone JF, Lazarus SA, Schmitz HH. Procyanidin content and variation in some commonly consumed foods. The Journal of Nutrition. 2000;**130**(8):2086S-2092S

[17] Ding EL, Hutfless SM, Ding X, Girotra S. Chocolate and prevention of

cardiovascular disease: A systematic review. Nutrition and Metabolism. 2006;**3**(1):2

[18] Wang C, Li S. Functional foods and nutraceuticals: Potential role in human health. In: Clinical Aspects of Functional Foods and Nutraceuticals. Boca Raton: CRC Press; 2014. pp. 72-97

[19] El Gharras H. Polyphenols: Food sources, properties and applications—A review. International Journal of Food Science & Technology. 2009;**44**(12):2512-2518

[20] Aziz AA, Edwards CA, Lean ME, Crozier A. Absorption and excretion of conjugated flavonols, including quercetin-4′-*O*-β-glucoside and isorhamnetin-4′-*O*-β-glucoside by human volunteers after the consumption of onions. Free Radical Research. 1998;**29**(3):257-269

[21] Crozier A, Lean ME, McDonald MS, Black C. Quantitative analysis of the flavonoid content of commercial tomatoes, onions, lettuce, and celery. Journal of Agricultural and Food Chemistry. 1997;**45**(3):590-595

[22] Anthony MS, Clarkson TB, Hughes CL Jr, Morgan TM, Burke GL. Soybean isoflavones improve cardiovascular risk factors without affecting the reproductive system of peripubertal rhesus monkeys. The Journal of Nutrition. 1996;**126**(1):43-50

[23] Burns J, Yokota T, Ashihara H, Lean ME, Crozier A. Plant foods and herbal sources of resveratrol. Journal of Agricultural and Food Chemistry. 2002;**50**(11):3337-3340

[24] Sanders TH, McMichael RW, Hendrix KW. Occurrence of resveratrol in edible peanuts. Journal of Agricultural and Food Chemistry. 2000;**48**(4):1243-1246

[25] Rubió L, Motilva MJ, Romero MP. Recent advances in biologically active compounds in herbs and spices: A review of the most effective antioxidant and antiinflammatory active principles. Critical Reviews in Food Science and Nutrition. 2013;**53**(9):943953

[26] Carocho M, Ferreira IC. The role of phenolic compounds in the fight against cancer—A review. Anti Cancer Agents in Medicinal Chemistry. 2013;**13**(8):1236-1258

[27] Cai YZ, Sun M, Xing J, Luo Q , Corke H. Structure–radical scavenging activity relationships of phenolic compounds from traditional Chinese medicinal plants. Life Sciences. 2006;**78**(25):2872-2888

[28] Carocho M, Ferreira IC. A review on antioxidants, prooxidants and related controversy: Natural and synthetic compounds, screening and analysis methodologies and future perspectives. Food and Chemical Toxicology. 2013;**51**:15-25

[29] Mizgier P, Kucharska AZ, Sokół-Łętowska A, Kolniak-Ostek J, Kidoń M, Fecka I. Characterization of phenolic compounds and antioxidant and anti-inflammatory properties of red cabbage and purple carrot extracts. Journal of Functional Foods. 2016;**21**:133146

[30] Bowen-Forbes CS, Zhang Y, Nair MG. Anthocyanin content, antioxidant, antiinflammatory and anticancer properties of blackberry and raspberry fruits. Journal of Food Composition and Analysis. 2010;**23**(6):554-560

[31] Jemal A, Bray F, Center MM, Ferlay J, Ward E, Forman D. Global cancer statistics. CA: A Cancer Journal for Clinicians. 2011;**61**(2):69-90

[32] Gordaliza M. Natural products as leads to anticancer drugs. Clinical and Translational Oncology. 2007;**9**(12):767-776

[33] Rice-Evans CA, Miller NJ, Paganga G. Structure-antioxidant activity

relationships of flavonoids and phenolic acids. Free Radical Biology and Medicine. 1996;**20**(7):933-956

[34] Tsang C, Higgins S, Duthie GG, Duthie SJ, Howie M, Mullen W, et al. The influence of moderate red wine consumption on antioxidant status and indices of oxidative stress associated with CHD in healthy volunteers. British Journal of Nutrition. 2005;**93**(2):233-240

[35] Carbonell-Capella JM, Buniowska M, Barba FJ, Esteve MJ, Frígola A. Analytical methods for determining bioavailability and bioaccessibility of bioactive compounds from fruits and vegetables: A review. Comprehensive Reviews in Food Science and Food Safety. 2014;**13**(2):155-171

[36] Rein MJ, Renouf M, Cruz-Hernandez C, Actis-Goretta L, Thakkar SK, da Silva Pinto M. Bioavailability of bioactive food compounds: A challenging journey to bioefficacy. British Journal of Clinical Pharmacology. 2013;**75**(3):588-602

[37] D'Archivio M, Filesi C, Varì R, Scazzocchio B, Masella R. Bioavailability of the polyphenols: Status and controversies. International Journal of Molecular Sciences. 2010;**11**(4):1321-1342

[38] Kawabata K, Mukai R, Ishisaka A. Quercetin and related polyphenols: New insights and implications for their bioactivity and bioavailability. Food & Function. 2015;**6**(5):13991417

[39] Wang Y, Wang X. Binding, stability, and antioxidant activity of quercetin with soy protein isolate particles. Food Chemistry. 2015;**188**:24-29

[40] Xiao Y, Lee IS. Microbial metabolism of prenylated apigenin derivatives by *Mucor hiemalis*. Phytochemistry Letters. 2016;**16**:197-202

[41] Velderrain-Rodríguez GR, Palafox-Carlos H, Wall-Medrano A, Ayala-Zavala JF, Chen CO, Robles-Sánchez M, et al. Phenolic compounds: Their journey after intake. Food & Function. 2014;**5**(2):189-197

[42] Kamonpatana K, Giusti MM, Chitchumroonchokchai C, MorenoCruz M, Riedl KM, Kumar P, et al. Susceptibility of anthocyanins to ex vivo degradation in human saliva. Food Chemistry. 2012;**135**(2):738-747

[43] Correa-Betanzo J, Allen-Vercoe E, McDonald J, Schroeter K, Corredig M, Paliyath G. Stability and biological activity of wild blueberry (*Vaccinium angustifolium*) polyphenols during simulated in vitro gastrointestinal digestion. Food Chemistry. 2014;**165**:522-531

[44] Zhao Z, Moghadasian MH. Bioavailability of hydroxycinnamates: A brief review of in vivo and in vitro studies. Phytochemistry Reviews. 2010;**9**(1):133-145

[45] Kobayashi S, Shinohara M, Nagai T, Konishi Y. Transport mechanisms for soy isoflavones and microbial metabolites dihydrogenistein and dihydrodaidzein across monolayers and membranes. Bioscience, Biotechnology, and Biochemistry. 2013;**77**(11):2210-2217

[46] Lewandowska U, Szewczyk K, Hrabec E, Janecka A, Gorlach S. Overview of metabolism and bioavailability enhancement of polyphenols. Journal of Agricultural and Food Chemistry. 2013;**61**(50):12183-12199

[47] Zhang X, Song J, Shi X, Miao S, Li Y, Wen A. Absorption and metabolism characteristics of rutin in Caco-2 cells. The Scientific World Journal. 2013;**2013**:1-8

[48] Bolli A, Marino M, Rimbach G, Fanali G, Fasano M, Ascenzi P.

Flavonoid binding to human serum albumin. Biochemical and Biophysical Research Communications. 2010;**398**(3):444-449

[49] Manach C, Scalbert A, Morand C, Rémésy C, Jiménez L. Polyphenols: Food sources and bioavailability. The American Journal of Clinical Nutrition. 2004;**79**(5):727-747

[50] Ajila CM, Rao UP. Mango peel dietary fibre: Composition and associated bound phenolics. Journal of Functional Foods. 2013;**5**(1):444-450

[51] Bordenave N, Hamaker BR, Ferruzzi MG. Nature and consequences of noncovalent interactions between flavonoids and macronutrients in foods. Food & Function. 2014;**5**(1):18-34

[52] Dona AM. Enhancing antioxidant activity and extractability of bioactive compounds of wheat bran using thermal treatments. Master [thesis]. University of Manitoba; 2011

[53] Sharma R. Polyphenols in health and disease: Practice and mechanisms of benefits. In: Polyphenols in Human Health and Disease. Cambridge, Massachusetts: Academic Press; 2014. pp. 757-778

[54] Pekkinen J, Rosa NN, Savolainen OI, Keski-Rahkonen P, Mykkänen H, Poutanen K, et al. Disintegration of wheat aleurone structure has an impact on the bioavailability of phenolic compounds and other phytochemicals as evidenced by altered urinary metabolite profile of diet-induced obese mice. Nutrition and Metabolism. 2014;**11**(1):1

[55] Renouf M, Marmet C, Giuffrida F, Lepage M, Barron D, Beaumont M, et al. Dose–response plasma appearance of coffee chlorogenic and phenolic acids in adults. Molecular Nutrition & Food Research. 2014;**58**(2):301-309

[56] Haslam E. Vegetable tannins—Lessons of a phytochemical lifetime. Phytochemistry. 2007;**68**(22-24):2713-2721

[57] Duval B, Shetty K, Thomas WH. Phenolic compounds and antioxidant properties in the snow alga *Chlamydomonas nivalis* after exposure to UV light. Journal of Applied Phycology. 1999;**11**(6):559

[58] Porrini M, Riso P. Factors influencing the bioavailability of antioxidants in foods: A critical appraisal. Nutrition Metabolism and Cardiovascular Diseases. 2008;**18**(10):647-650

[59] Hamauzu Y, Inno T, Kume C, Irie M, Hiramatsu K. Antioxidant and antiulcerative properties of phenolics from Chinese quince, quince, and apple fruits. Journal of Agricultural and Food Chemistry. 2006;**54**(3):765-772

[60] Grijalva EG, Grijalva PG, Pérez DLA, López NL, López RIC, Heredia JB. Bioavailability of dietary phenolic compounds. Revista española de nutrición humana y dietética. 2016;**20**(2):140-147

[61] Parkinson A, Ogilvie BW. Biotransformation of xenobiotics. In: Klaassen CD, Watkins JB, editors. Casarett & Doull's Essentials of Toxicology. McGraw-Hill Medical; 2010

[62] Karakaya S. Bioavailability of phenolic compounds. Critical Reviews in Food Science and Nutrition. 2004;**44**(6):453-464

[63] Scalbert A, Williamson G. Dietary intake and bioavailability of polyphenols. The Journal of Nutrition. 2000;**130**(8):2073S-2085S

[64] Crozier A, Del Rio D, Clifford MN. Bioavailability of dietary flavonoids and phenolic compounds. Molecular Aspects of Medicine. 2010;**31**(6):446-467

[65] Poquet L, Clifford MN, Williamson G. Effect of dihydrocaffeic acid on UV irradiation of human keratinocyte HaCaT cells. Archives of Biochemistry and Biophysics. 2008;**476**(2):196-204

[66] Basheer L, Kerem Z. Interactions between CYP3A4 and dietary polyphenols. Oxidative Medicine and Cellular Longevity. 2015;**2015**:1-15

[67] Misaka S, Kawabe K, Onoue S, Werba JP, Giroli M, Watanabe H, et al. Green tea extract affects the cytochrome P450 3A activity and pharmacokinetics of simvastatin in rats. Drug Metabolism and Pharmacokinetics. 2013;**28**(6):514-518

[68] Aura AM. Microbial metabolism of dietary phenolic compounds in the colon. Phytochemistry Reviews. 2008;7(3):407-429

[69] Roowi S, Stalmach A, Mullen W, Lean ME, Edwards CA, Crozier A. Green tea flavan-3-ols: Colonic degradation and urinary excretion of catabolites by humans. Journal of Agricultural and Food Chemistry. 2009;**58**(2):1296-1304

[70] Touriño S, Fuguet E, Vinardell MP, Cascante M, Torres JL. Phenolic metabolites of grape antioxidant dietary fiber in rat urine. Journal of Agricultural and Food Chemistry. 2009;**57**(23):11418-11426

[71] Selma MV, Espin JC, Tomas-Barberan FA. Interaction between phenolics and gut microbiota: Role in human health. Journal of Agricultural and Food Chemistry. 2009;**57**(15):6485-6501

[72] Kroon PA, Clifford MN, Crozier A, Day AJ, Donovan JL, Manach C, et al. How should we assess the effects of exposure to dietary polyphenols in vitro? The American Journal of Clinical Nutrition. 2004;**80**(1):15-21

[73] Burchell B, Leakey J, Dutton GJ. Relationship between activation of 'Detoxicating' enzymes in stored broken-cell preparations and in autolysing liver. Enzyme. 1975;**20**:156-164

[74] Hawes EM. N+-Glucuronidation, a Common Pathway in Human Metabolism of Drugs with a Tertiary Amine Group: 1996 ASPET N-Glucuronidation of Xenobiotics Symposium. 1998. p. 830-837

[75] Del Rio D, Costa LG, Lean MEJ, Crozier A. Polyphenols and health: What compounds are involved? Nutrition, Metabolism, and Cardiovascular Diseases. 2010;**20**(1):1-6

[76] Steffen Y, Gruber C, Schewe T, Sies H. Mono-*O*-methylated flavanols and other flavonoids as inhibitors of endothelial NADPH oxidase. Archives of Biochemistry and Biophysics. 2008;**469**(2):209-219

[77] Actis-Goretta L, Lévèques A, Rein M, Teml A, Schäfer C, Hofmann U, et al. Intestinal absorption, metabolism, and excretion of (−)-epicatechin in healthy humans assessed by using an intestinal perfusion technique. The American Journal of Clinical Nutrition. 2013;**98**(4):924-933

[78] Spencer JP, Schroeter H, Rechner AR, Rice-Evans C. Bioavailability of flavan3-ols and procyanidins: Gastrointestinal tract influences and their relevance to bioactive forms in vivo. Antioxidants and Redox Signaling. 2001;**3**(6):1023-1039

[79] Das S, Rosazza JP. Microbial and enzymatic transformations of flavonoids. Journal of Natural Products. 2006;**69**(3):499-508

[80] Bernini R, Crisante F, Ginnasi MC. A convenient and safe *O*-methylation of flavonoids with dimethyl carbonate (DMC). Molecules. 2011;**16**(2):1418-1425

[81] van Zanden JJ, Wortelboer HM, Bijlsma S, Punt A, Usta M, van Bladeren PJ, et al. Quantitative structure activity relationship studies on the flavonoid mediated inhibition of multidrug resistance proteins 1 and 2. Biochemical Pharmacology. 2005;**69**(4):699-708

[82] Walle T. Methylation of dietary flavones increases their metabolic stability and chemopreventive effects. International Journal of Molecular Sciences. 2009;**10**(11):5002-5019

[83] Caleja C, Ribeiro A, Filomena Barreiro M, CFR Ferreira I. Phenolic compounds as nutraceuticals or functional food ingredients. Current Pharmaceutical Design. 2017;**23**(19):2787-2806

[84] Dias MI, Ferreira IC, Barreiro MF. Microencapsulation of bioactives for food applications. Food & Function. 2015;**6**(4):1035-1052

[85] Li W, Friel J, Beta T. An evaluation of the antioxidant properties and aroma quality of infant cereals. Food Chemistry. 2010;**121**(4):1095-1102

[86] Bononi M, Tateo F. Stabilization of cranberry anthocyanins in nutraceutical capsules. International Journal of Food Sciences and Nutrition. 2007;**58**(2):142-149

[87] Martins A, Barros L, Carvalho AM, Santos-Buelga C, Fernandes IP, Barreiro F, et al. Phenolic extracts of *Rubus ulmifolius* Schott flowers: Characterization, microencapsulation and incorporation into yogurts as nutraceutical sources. Food & Function. 2014;**5**(6):1091-1100

[88] Caleja C, Barros L, Antonio AL, Carocho M, Oliveira MBP, Ferreira IC. Fortification of yogurts with different antioxidant preservatives: A comparative study between natural and synthetic additives. Food Chemistry. 2016;**210**:262-268

[89] Caleja C, Barros L, Antonio AL, Ciric A, Barreira JC, Sokovic M, et al. Development of a functional dairy food: Exploring bioactive and preservation effects of chamomile (*Matricaria recutita* L.). Journal of Functional Foods. 2015;**16**:114-124

[90] Caleja C, Barros L, Antonio AL, Ciric A, Soković M, Oliveira MBP, et al. *Foeniculum vulgare* mill. as natural conservation enhancer and health promoter by incorporation in cottage cheese. Journal of Functional Foods. 2015;**12**:428-438

[91] Ezhilarasi PN, Indrani D, Jena BS, Anandharamakrishnan C. Freeze drying technique for microencapsulation of *Garcinia* fruit extract and its effect on bread quality. Journal of Food Engineering. 2013;**117**(4):513-520

[92] Chouchouli V, Kalogeropoulos N, Konteles SJ, Karvela E, Makris DP, Karathanos VT. Fortification of yoghurts with grape (*Vitis vinifera*) seed extracts. LWT—Food Science and Technology. 2013;**53**(2):522-529

[93] Pasrija D, Ezhilarasi PN, Indrani D, Anandharamakrishnan C. Microencapsulation of green tea polyphenols and its effect on incorporated bread quality. LWT—Food Science and Technology. 2015;**64**(1):289-296

[94] Alves G, Perrone D. Breads enriched with guava flour as a tool for studying the incorporation of phenolic compounds in bread melanoidins. Food Chemistry. 2015;**185**:65-74

[95] Pillai DS, Prabhasankar P, Jena BS, Anandharamakrishnan C. Microencapsulation of *Garcinia cowa* fruit extract and effect of its use on pasta process and quality. International

Journal of Food Properties. 2012;**15**(3):590-604

[96] Robert P, Gorena T, Romero N, Sepulveda E, Chavez J, Saenz C. Encapsulation of polyphenols and anthocyanins from pomegranate (*Punica granatum*) by spray drying. International Journal of Food Science & Technology. 2010;**45**(7):1386-1394

[97] Çam M, İçyer NC, Erdoğan F. Pomegranate peel phenolics: Microencapsulation, storage stability and potential ingredient for functional food development. LWT—Food Science and Technology. 2014;**55**(1):117-123

[98] Ribeiro A, Ruphuy G, Lopes JC, Dias MM, Barros L, Barreiro F, et al. Spray-drying microencapsulation of synergistic antioxidant mushroom extracts and their use as functional food ingredients. Food Chemistry. 2015;**188**:612-618

[99] Stojkovic DS, Zivković J, Sokovic M, Glamoclija J, Ferreira IC, Jankovic T, et al. Antibacterial activity of *Veronica montana* L. extract and of protocatechuic acid incorporated in a food system. Food and Chemical Toxicology. 2013;**55**:209-213

[100] Fan P, Lou H. Effects of polyphenols from grape seeds on oxidative damage to cellular DNA. Molecular and Cellular Biochemistry. 2004;**267**(1-2):67-74

[101] Stagos D, Spanou C, Margariti M, Stathopoulos C, Mamuris Z, Kazantzoglou G, et al. Cytogenetic effects of grape extracts (*Vitis vinifera*) and polyphenols on mitomycin C-induced sister chromatid exchanges (SCEs) in human blood lymphocytes. Journal of Agricultural and Food Chemistry. 2007;**55**(13):5246-5252

[102] Ugartondo V, Mitjans M, Lozano C, Torres JL, Vinardell MP. Comparative study of the cytotoxicity induced by antioxidant epicatechin conjugates obtained from grape. Journal of Agricultural and Food Chemistry. 2006;**54**(18):6945-6950

[103] Fu L, Xu BT, Xu XR, Gan RY, Zhang Y, Xia EQ, et al. Antioxidant capacities and total phenolic contents of 62 fruits. Food Chemistry. 2011;**129**(2):345-350

Chapter 6

Polyphenols and Athletic Performance: A Review on Human Data

Stefania D'Angelo

Abstract

Exercise-induced aerobic bioenergetic reactions in mitochondria and cytosol increase production of reactive oxygen species. Many efforts have been carried out to identify dietary strategies or micronutrients able to prevent or at least attenuate the exercise-induced muscle damage and stress. A lot of studies are about how effective dietary intervention and oral antioxidant supplementation may be in reducing oxidative stress in athletes who exercise intensively. Commonly used nonenzymatic supplements have been proposed as ways to prevent exercise-induced oxidative stress and hence improve adaptation responses to endurance training. Plant-derived bioactive compounds can repress inflammation by inhibiting oxidative damage and interacting with the immune system. This review focuses on polyphenols and phytochemicals present in the plant kingdom that have been recently suggested to exert some positive effects on exercise-induced muscle damage and oxidative stress. This review will summarize some of the actual knowledge on polyphenolic compounds that have been demonstrated both to exert a significant effect in exercise-induced muscle damage and to play a biological/physiological role in improving physical performance. Overall, the pooled results show that polyphenols are viable supplements to improve performance in athletes.

Keywords: antioxidant, athletic performance, nutrition, polyphenols, oxidative stress, sport

1. Introduction

Many authors have noted that physical exercise induces an increase in production of free radicals and other reactive oxygen species (ROS) [1]. Current evidence indicates that ROS are the primary reason of exercise-induced disturbances in muscle redox balance, and it was observed that severe disturbances in redox balance have been shown to promote oxidative injury and muscle fatigue impairing the exercise performance [2, 3].

During the physical exercise practice, it is possible for the activation of some biological processes, such as the macrophages infiltration, the movement of electrons that occurs at the level of the transport chain on the mitochondrial ridges, the catabolism pathway of the purines, or the reaction catalyzed by the enzyme xanthine oxidase, which all may lead to the release of ROS. On the basis of the above-mentioned information, sportsmen have to improve their antioxidant defense systems to overcome the exercise-induced oxidative damage. It is well established that the increase in reactive oxygen species and free radical production during exercise has both positive and negative physiological effects. In 2008, for the first time, moderate exercise has been defined as an antioxidant, explaining that the mild burst of ROS, generated by training, acts as a signal responsible for the activation of signaling pathways that lead to the induction of antioxidant enzymes in human tissue [4]. To prevent these hypothetically negative or side effects of physical exercise, supplementation with different types of antioxidants has been used in a great number of studies [2].

The term "antioxidant" in general indicates the molecules capable of preventing, delaying or, in some cases, completely canceling oxidative damage to specific target molecules. For example, the superoxide dismutase enzyme, the catalase enzyme, and the glutathione peroxidase enzyme are endogenous antioxidants; glutathione, vitamins E, C and A, and coenzyme Q10 (CoQ10) are nonenzymatic molecules with antioxidant properties [5].

The antioxidants can also be taken through the "exogenous antioxidant" diet, and this supplementation can improve the ability to protect muscle fiber, during training, from oxidative damage caused by fatigue. In fact, the deficiency of antioxidants could induce an increased predisposition to oxidative damage induced by exercise and therefore compromise the sporting performance [6].

Over the past few decades, many attempts have been made to improve antioxidant potential, and therefore increase physical performance by improving nutrition, training programmers, and other related factors. Recently, the problem of whether or not athletes should use antioxidant supplements is an important and highly debated topic.

In the context of this chapter, information in brief about the well-known and recently used antioxidants in particular the polyphenols is given. This review describes only human trials. The effects of these antioxidants on exercise performance and exercise-induced oxidative stress are also explained.

2. Oxidative stress and antioxidants

All aerobic organisms constantly synthesize free radicals as part of normal metabolic processes. Free radicals are chemical species with an unpaired electron in their outermost orbital; the free radicals that can be formed precisely by the oxygen molecules are indicated by the acronym ROS, that is, reactive oxygen species [7, 8].

Free radicals and reactive oxygen species are the main oxidizing agents in cellular systems and are involved in aging and the onset of many types of diseases. They are physiologically produced in different cellular biochemical reactions occurring in the body, such as in mitochondria for aerobic oxygen production, in fatty acid metabolism, in drug metabolism, and during activity of the immune system. On the other hand, free radicals can also be produced by exogenous factors such as pollution, bad lifestyle habits, UV rays, ionizing radiation, and psychophysical stress resulting from intense physical activity [9–11]. Although these free radicals have positive effects on immune reactions and cellular signaling, they are also known to have negative effects, such as oxidative damage of lipids, proteins, and nucleic acids. Organisms are equipped with antioxidant defense systems that protect cells from the toxic effects of free radicals [9–11].

Antioxidants are molecules able to give an electron to free radicals, neutralizing, diminishing, or eliminating their ability to damage cells and the main biomolecules such as nucleic acids, proteins, and lipids.

As already mentioned, it is possible to divide the antioxidants into two categories: enzymatic antioxidants, such as the enzyme superoxide dismutase (SOD), the enzyme glutathione peroxidase, or the enzyme catalase; nonenzymatic antioxidants, such as glutathione, vitamin E, vitamin C, and bilirubin.

These "endogenous" antioxidants have the function of delaying or preventing the oxidation of extracellular and intracellular biomolecules. We know that even antioxidants taken from the diet, such as vitamins and minerals, can condition the oxidative state of the body. Some mammals, except humans, possess the biochemical mechanism that allows them to synthesize vitamin C. And this information can be useful for the purpose of supplementary integration during a sports performance [9].

Therefore, oxidative stress produces oxidative damage that can influence various physiological functions and can be defined as an imbalance between oxidants and antioxidants in favor of oxidants.

The production of ROS induced by exercise is an important signaling path for inducing biological adaptations to training, but ROS production could also have a deleterious impact on cells and tissues, that is, lipid and protein peroxidation. This concern has led some experts to suggest consuming more dietary supplements and supplements containing antioxidant to mitigate ROS production which can cause excessive oxidative stress during and after exercise [9–12].

2.1 Oxidative stress and physic activity

We have said that free radicals are normally generated during various physiological mechanisms. But their production increases considerably during a physical activity; in this situation, the skeletal muscles need a greater oxygen supply, resulting in an obvious change in the blood flow between the various organs. Subsequently, muscle damage induced by physical exercise causes infiltration of phagocytes (macrophages and neutrophils) in the area where the lesion occurred. All these physiological changes that occur during acute exercise cause an increase in the production of ROS, with consequent oxidative damage to the biomolecules. Through the use of biochemical and molecular techniques, it is now possible to evaluate events that occur at the cellular level and demonstrate in an increasingly precise manner, as free radicals certainly play a role in the physiological adaptations observed in the athlete after training. But free radicals generated by exercise can have both positive and negative physiological effects [10].

Exercise-induced oxidative stress associated with increased free radical production has been studied for 40 years, since it was first reported in 1978. In the study, subjects perform a 60-min cycle ergometer exercise at 50% VO_2 max intensity and reported increased levels of expired pentane, an index of lipid peroxidation [10]. Subsequently, in 1987, a study was carried out on six young men who performed an incremental load exercise on a cycle ergometer until it was exhausted, and it was discovered that the blood levels of reactive substances to thiobarbituric acid (TBARS), another marker of lipid peroxidation, increased [24]. In another study, in 1988, in which eight highly trained young men performed cycle ergometer exercise for 90 min at 65% VO_2 peak intensity, the levels of GSH, a nonenzymatic antioxidant, decreased, whereas the GSSG levels conversely increased [10].

The expression of the proteins involved in mitochondrial biogenesis, that is, the receptor gamma-activator receptor activated by the peroxisome alpha proliferator (PGC-1α) is increased by regular resistance training. In fact, ROS stimulates the cascade of mitochondrial biogenesis precisely in response to endurance training, that is, the chronic muscular contractions [12]. The newly formed mitochondria are highly efficient and have the capacity to synthesize less ROS for the same amount of adenosine triphosphate product, that is, they are more functional.

For example, expression of PGC-1α in skeletal muscle was significantly increased following 4 weeks of endurance training [13], indicating a skeletal muscle contraction-stimulated mechanism of mitochondrial biogenesis. During muscle contraction, ROS can also be generated through mechanisms that do not involve the mitochondria. In fact, it has been shown that muscle contraction causes an increase in superoxide ion in the cytosol, before the increase in mitochondria occur. It has been hypothesized that the activity of nicotinamide adenine dinucleotide phosphate-oxidase (NADPH) causes the superoxide ion to increase in the cytosol [14]. Accordingly, ROS production (level of H_2O_2) was previously shown to increase in isolated mitochondria after acute muscle contraction in comparison with rested skeletal muscle biopsy sample [10].

ROS production leads to muscle fiber damage, which eventually results in muscle fatigue.

ROS production leads to damage to the muscle fiber, which then results in muscle strain. However, increasing evidence suggests that small stimuli, such as a low concentration of ROS, are able to stimulate the transcription of the major genes that encode proteins with antioxidant power. Superoxide dismutase (SOD) and glutathione are the examples of important molecules capable of defending cells from ROS-induced oxidative stress. Being able to deepen the mechanism of correlation between muscle fatigue and oxidative damage could be an important strategy for nutritional interventions aimed at increasing the performance of the exercise. An effective strategy could be antioxidant supplementation, considering the effects of scavenging ROS, which could lead to a decrease in muscle damage caused by prolonged exercise [14].

To date, there has been a plethora of reports on the effects of acute aerobic exercise on oxidative stress markers. Most of these studies have been conducted on young healthy men, and the difficulty of linking the various studies depends on the fact that, in each study, the training status is different. The most common exercises are those in which the athletes analyzed use a cycle ergometer or treadmill, in which the subjects, in an air-conditioned laboratory, generally engage in a maximal or submaximal exercise for 10–90 min.

Other studies have evaluated the effects of eccentric contraction exercises, such as downhill exercises. The most analyzed biological sample is blood. In a limited number of studies, skeletal muscle, exhaled air, and urine were also examined. Oxidation products of lipids, proteins, and DNA (i.e., MDA, PC, and 8-OHdG) have been used as oxidative stress markers; antioxidant levels and the redox balance in tissues were also assessed [3, 8].

Oxidative stress can be induced by aerobic exercise, but it can also be activated by anaerobic exercise. In fact, in addition to studies using sprint exercises, other studies have evaluated the effects of resistance exercises (formation of muscle groups throughout the body through the use of different types of resistance exercises) on oxidative stress markers. In a study in which 12 highly trained youths performed three sets of eight types of resistance exercise at 10 repetition loads, levels of malondialdehyde, a marker of lipid peroxidation, increased in the blood [8, 9]. This is one of the many tests that demonstrate that resistance exercises involving the whole musculature of the body modify the blood levels of oxidative stress markers [8]. But also local resistance exercises (defined as exercises that train a specific muscle group using a single type of resistance exercise) modify the oxidative stress markers of the blood [8]. While all these studies have assessed oxidative stress by measuring the change in blood parameters, other studies have observed muscle biopsies, to demonstrate that local resistance exercises increase oxidative stress in skeletal muscles [8].

On the other hand, other studies have shown that levels of oxidative stress in the blood are not conditioned by resistance exercises. Probably, the discrepancy of this data may have been conditioned by the training status. However, a study of individual changes in oxidative stress responses to eccentric exercise demonstrated high inter-individual variability after exercise of eccentric knee extension, even in subjects with the same training status [8]. Furthermore, this study showed that in about one person in three exercise-induced oxidative stress was unexpected or negligible (response rate of 5% or less). These data reasonably suggest that the inconsistencies between the various results, both for the anaerobic and aerobic exercises, can be caused both by the training status, but also by the great inter-individual variability of the response to the oxidative stress induced by physical activity [3, 8].

2.2 Antioxidants and exercise

An active debate still exists on the effect of antioxidant supplementation on exercise-induced oxidative stress. Typical treatment generally includes vitamins A, C, and E, at various dosages, administered alone or in combination, chronically or acutely [15, 16]. Of these, vitamins C and E have been used more frequently in clinical and experimental studies, mostly because of their safety profile and easy availability [15]. One study showed the administration of vitamin C (500 mg, a moderate dose), reduced exercise-induced lipid peroxidation, and muscle damage in an untrained male group compared to the placebo group; on the contrary, it had no effect on inflammatory markers [17].

Other less-used antioxidants include coenzyme Q10 and N-acetylcysteine [15].

Regarding the endpoints, it is possible to hypothesize that the antioxidant could be effective in particular conditions in terms of training; for example, a specific moment of training (for example, before or after the race) or a type of sport compared to another (e.g., anaerobic versus aerobic).

Therefore, to decide whether to administer an antioxidant supplement, the selection and detailed description of the appropriate training stimulus and/or monitoring of the athlete during the training phases is necessary [15].

However, it is important to underline that numerous studies report negative effects of antioxidants. It is hypothesized that one of the motivations for the controversy is the different population analyzed in the studies.

In most of the studies in which a benefit of antioxidant supplementation was demonstrated in attenuating muscle damage and oxidative stress following endurance exercise, data on samples of sedentary and nonresistant subjects were analyzed. The endogenous antioxidant defenses of trained subjects could be over regulated and therefore these subjects may not benefit greatly from the use of exogenous antioxidants [15]. Furthermore, another big difference is represented by the different types of exercise; in fact, we move from the exercise of short-term resistance to the exercise of long-term resistance. In fact, aerobic endurance exercise will surely induce, due to the massive use of oxygen, a different flow of radicals with respect to the exercise of anaerobic resistance [15].

Recently, a large body of literature has highlighted a potential relationship between oxidative stress and the bioactive compounds present in plant foods. In particular, the researchers' attention has been shifted to the effects of a peculiar class of bioactive nutraceutical compounds, that is, polyphenols.

The study of phenolic compounds present in food has attracted great interest since the 1990s due to the growing evidence of their beneficial effect on human health. One of the first studies that stimulated the interest of scientists is the epidemiological study of Zutphen. In this research, an inverse association was proposed between the intake of foods rich in polyphenols and the incidence of diseases, such as diabetes mellitus, cardiovascular diseases, and cancer [18] and in particular of those pathologies associated with an evident oxidative stress.

Therefore, efforts to develop dietary strategies against oxidative stress caused by physical activity are being made and recently, there has been a growing interest in investigating the potential of polyphenols to modulate physical performance and prevent oxidative stress.

3. Polyphenols

Human diets are rich in polyphenols; western populations consume an estimated 1–2 g/day polyphenols, mainly from fruits, vegetables, and beverages such as tea, coffee, wine, and fruit juices. Polyphenols exert a range of biological activities, and various epidemiological studies and clinical trials have linked their intake with a reduced risk of chronic diseases, such as coronary heart disease, stroke, type II diabetes, and some cancers. There has recently been growing interest, supported by a number of epidemiological and experimental studies, on the possible beneficial effects of polyphenols on brain health.

For this reason, these phytochemicals are currently considered important components of a healthy diet, and it is believed that the health benefits of a diet rich in fruit and vegetables are to be attributed to these molecules. For example, the protective effects of tea against cardiovascular disease or coffee against type II diabetes could be explained by the large concentration of catechins present in these drinks.

As a consequence, the scientific and commercial interest in these phytochemicals has increased considerably in recent years; in particular, there are numerous works in which topics concerning their bioavailability, bioactivity, metabolism, and health effects have been addressed [19–21].

3.1 Nomenclature, classifications, and occurrence in foods

Polyphenols are classified into flavonoids and nonflavonoids, according to the number of phenol rings and structural elements bound to these rings.

Flavonoids represent the largest group of polyphenols. The chemical structure is characterized by 15 carbon atoms, derived from the flavone and all share certain properties. They are mainly soluble in water; usually, they are present in the plant as glycosides and in the same plant, there can be a flavonoid aglycone in combination with different sugars. Their name derives from flavus (=yellow) and refers to the role they play as plant pigments. The coloring that they give to the tissues depends on the pH. A specific group of flavonoids, the anthocyanins, is responsible for the red, blue, and violet colors of flowers and fruit and is therefore very important as a mediator of pollination. It is therefore not surprising that the variety of shades of color associated with anthocyanins has been increasing through the evolutionary process.

There are six dietary groups of flavonoids:

- flavones (luteolin and apigenin), which are found in parsley, capsicum pepper, and celery;

- flavonols (kaempferol, quercetin, and myricetin), which are found in onions, leeks and broccoli, cherry tomato, apple, berries, beans, tea and red wine;

- flavanones (hesperetin, naringenin, and eriodictyol), mainly present in orange, lemon juice, grapefruit, and herbs (oregano);

- isoflavones (genistein and daidzein), which are mainly found in soybeans;

- flavanols (catechin, epicatechin, and gallocatechin), abundant in green tea, apple, red wine, cocoa, chocolate and may be present as monomers or as oligomers; in particular flavan-3,4-diol polymerization produces so-called "condensed tannins": they are also the origin of the catechins and probably polymerize with them to give the proanthocyanidins;

- anthocyanidins or anthocyanins (delphinidin, pelargonidin, cyanidin, malvidin, and petunidin), whose sources include red wine, berries, blackcurrant, and cherry [22–25].

The nonflavonoid group can be separated into three different classes: phenolic acids, stilbenes, and lignans.

Phenolic acids can be found in many plant species, in dried fruit. The most common phenolic acid are:

- caffeic acid is generally the most abundant phenolic acid and is mainly found as a quinic ester; present in many fruits and vegetables; it is a major phenolic compound in coffee;

- chlorogenic acid is the ester of caffeic acid and quinic acid; it is present in blueberries, kiwis, prunes, and apples;

- ferulic acid present in cereals, which is esterified to hemicellulose in the cell wall.

The best studied stilbene is resveratrol, and it can be found in *cis* or *trans*, or glucosylate, or in lower concentrations as the parent molecule of a family of polymers such as the food sources of viniferine, pallidol, or ampelopsin A. The resveratrol is found in particular in the grape skin and in the wine, to a greater extent in the red one.

Lignans (secoisolariciresinol, matairesinol, medioresinol, pinoresinol, and lariciresinol) are found in high concentration in linseed and in minor concentration in algae, leguminous plants, cereals, vegetables, and fruits [22–24].

To the large group of polyphenols are also added some molecules that form a separate category, compared to the four high. These include tyrosol and curcumin, which have long been discussed for potential health benefits.

The main classes of polyphenols present in the common diet are the flavanols, in particular the catechins and tannins of tea; the flavanones, mostly hesperidin present in citrus fruits; flavonols, such as quercetin in tea, apples, and onions; hydroxycinnamic acids, phenolic acids, abundant in coffee and many fruits and vegetables; anthocyanins, polyphenols responsible for the color of fruit and vegetables.

Among others, fruits like apples, grapes, pears, and berries typically contain high amounts of polyphenols (200–300 mg/100 g). Other polyphenol compound is curcumin (1,7-bis-(4-hydroxy-3-methoxyphenyl)-1,6-heptadiene-3,5-dione), a member of the curcuminoid family. It is found in turmeric, a spice produced from the rhizome of Curcuma long. Moreover, olive oil is a source of at least 30 phenolic compounds; the three phenolic compounds found in the highest concentration in olive oil are oleuropein, hydroxytyrosol, and tyrosol [21, 23].

Since agriculture was developed in 10,000 bc, humans have been modifying plant secondary metabolite profiles. By selecting fruit, flower, and vegetable colors, farmers involuntarily elected higher anthocyanin content, whereas in selecting for scents, they modified volatile phenolics. Anyway, as regards the content of polyphenols present in foods, it seems impossible to classify basic foods in terms of "how many polyphenols they provide annually". However, the most important food products, consumed in large quantities, which are the main source of polyphenols, are fruits and vegetables, green and black tea, red wine, coffee, chocolate, and extra virgin olive oil.

Even herbs and spices, nuts and algae are possible sources of food polyphenols; this depends on the culinary traditions and habits.

In 2010, a website was created (www.phenol-explorer.eu) containing over 35,000 content values for 500 different polyphenols in over 400 foods [16, 26].

Numerous factors can affect the content of polyphenols in food, may be environmental factors, such as exposure to the sun, precipitation, different types of crops, different types of soil, yield in fruits for the tree, degree of ripeness, but also conservation and methods of culinary preparation [27, 28].

3.2 Bioavailability of polyphenols

Bioavailability is defined as the fraction of a nutrient that the body is able to absorb and use for its physiological functions. Bioavailability may vary in relation to numerous factors, depending in part on the nature of the food and partly on the characteristics of the organism that assumes it. It is evident that bioavailability is very important in the nutritional field, but often neglected. A compound could have strong antioxidant activities or other biological activities *in vitro*, but if it were not bioavailable, if only a small amount of this type reached the target tissues, the molecule would have little biological activity *in vivo*. It is therefore essential to estimate the bioavailability of the polyphenols in order to set up effective nutritional protocols.

There are numerous studies carried out *in vitro*, then performed using as experimental models cultured cells or tissue slices, which provided fundamental information to understand the beneficial effects of polyphenols.

Despite the encouraging results, it is advisable to be very cautious in interpreting and extracting the data obtained in these experiments. Indeed, aglycons are often tested instead of active metabolites. Many researchers, when trying to unravel the physiological mechanisms involved in the health effects of polyphenols, often analyze the properties of a compound of little biological relevance. The dose used is also important, which should, in reality, be approximated to real life conditions. *In vitro* concentrations commonly range from low μmol/L to mmol/L, while plasma metabolite concentrations, after a normal dietary intake, rarely exceed nmol/L. Using excessive doses in experiments performed *in vitro* could "force" positive outcomes; therefore, the results obtained must be extrapolated with great care in *in vivo* situations [23, 29].

3.3 Biological effects

Historically, polyphenols were mostly of interest to botanists, as they play many roles in plants. In the 1990s, polyphenols were classified as general antioxidants [30], and it was thought that it was easy to explain their activity. The reality of polyphenols is much more complex: biological effects involve detailed biochemical interactions with pathways at the molecular level. Much progress has been made in the last few decades.

Although the chemical structure gives polyphenols primarily an antioxidant activity, this property does not necessarily represent their biological effect, since any action on the organism depends on both bioavailability and molecular targets.

The overall effect of these phytochemicals on the reduction of disease risk is supported by epidemiology, where foods and beverages rich in polyphenols are protective against the development of certain chronic diseases, in particular cardiovascular diseases, type 2 diabetes, and cancer [21, 31–35].

In recent years, revisions have been published supporting a new effect of polyphenols, that is, their role as supplements in athletic performance [11, 12].

However, despite these reviews, the overall effect of polyphenols on performance is inconclusive as most studies involve a small sample size.

Polyphenols are natural antioxidants present in the human diet in which they can reduce the damage due to the production of ROS. The chemical characteristics of the polyphenols allow them to act as direct scavengers of free radicals, such as the catechol group on the B ring, the presence of hydroxyl groups on the 3 and 5 position, and the 2,3-double bond in conjugation with a 4-oxofunction of a carbonyl group in the C-ring [9]. However, polyphenols can also behave as pro-oxidants at high doses or in the presence of metal ions, leading to DNA degradation. There is no evidence of systemic pro-oxidant effect of polyphenols in humans [21, 36]. Because of the low bioavailability and the kinetic constraints, the direct antioxidant activity of the polyphenols appears to be ineffective *in vivo*. Therefore, it has been hypothesized that the beneficial effects are not due to the direct scavenger properties on ROS, but to an indirect antioxidant action. In fact, polyphenols can modulate gene expression by inducing the endogenous antioxidant enzyme defense system.

4. Polyphenols action on athletic performance

As of today, a lot of studies are on polyphenols and physical exercise and in particular the polyphenolic compounds that have been demonstrated both to exert a significant effect in exercise-induced muscle damage and to play a biological/physiological role in improving physical performance. The effects of different polyphenols have been investigated in a wide range of exercise conditions, using a variety of supplementation strategies, timing, and dosage. Until a few years ago, despite the active search for "natural," polyphenol-rich extracts that might enhance physical performance and decrease oxidative damage because they are antioxidants, the information we had was very limited and in some cases, they suggested the converse [37]. But in recent years, studies have increased considerably, and more information is now available on the effect of polyphenols on sports performance [14, 38–42].

4.1 Quercetin

Among nutraceutical compounds, flavonoids are the mainly studied ones for their positive effects on human health, and some of them have been proposed to be beneficial in exercise and exercise performance. Among flavonols, quercetin accounts for about 13.82 mg/day, resulting in being one of the most abundant flavonols in Western diet. Quercetin (3,4,5,7-pentahydroxylflavone) is a natural bioactive flavonoid, mainly present as quercetin glycosides (rutin, spiraeoside, troxerutin, quercitrin, isoquercetin, and hyperoside). It is distributed in a wide variety of natural foods, such as nuts, grapes, broccoli, and black tea; it is found in apples, berries, onions, grapes, and tomatoes as well as in some medicinal plants as *Hypericum perforatum* and Gingko biloba [42].

Antioxidant properties of quercetin are attributed to its chemical structure, particularly the presence and location of the hydroxyl (–OH) substitutions. Quercetin supplementation studies in athletes have focused on the potential effects of exercise-induced inflammation, oxidative stress, immune dysfunction, and exercise performance [42]. The first human exercise study investigating quercetin supplementation was published in 2006, with many more published in the past few years and continuing to be published. When athletes are studied, most of the researches have failed to find an ergogenic effect, in contrast to that of a study of elite cyclists, who exhibited an improvement of their aerobic performance, and another study indicated that administration of quercetin (1200 mg) for 6 weeks resulted in performance improvement in cyclists [43].

The effects of quercetin supplementation in cycling athletes have been investigated [44]. Forty athletes were recruited and randomized to quercetin or placebo. Subjects consumed 1000 mg quercetin or placebo each day for 6 weeks before and during 3 days of cycling at 57% work maximum for 3 h. Despite previous data demonstrating potent antioxidant actions of quercetin in *in vitro* and animal models, long-term quercetin supplementation was not able to exert any preventive effect on exercise-induced oxidative stress and inflammation biomarkers. In another study, the influence of 1000 mg quercetin with or without 120 mg of epigallocatechin 3-gallate, 400 mg of isoquercetin and 400 mg of eicosapentaenoic acid and docosahexaenoic acid was evaluated on sports performance, biogenesis of muscle mitochondria and changes of markers of immunity and inflammation before and after a 3-day period of heavy effort. Two-week supplementation with polyphenols was effective in augmenting inflammation after 3 days of heavy exertion in trained cyclists. The feeding of untrained healthy men and women was supplemented with 1000 mg quercetin for 7 days, and the effect on VO_2 max and fatigue time was evaluated using a bicycle ergometer. Both fatigue (13.2%) and VO_2 max (3.9%) increases were found [42].

Other studies have analyzed the effect of quercetin on exercise performance, some reporting positive effects, while others do not, but to our knowledge, an increase in mitochondrial biogenesis has not been reported in human even though it has been shown a modest and insignificant increase in relative mitochondrial DNA copy number following quercetin supplementation [45].

A meta-analysis results have demonstrated that polyphenol supplementation for at least 7 days increases performance by 1.90%. Sub-analysis of seven studies using quercetin identified a performance increase of 2.82% [41]. There were no adverse effects reported in the studies in relation to the intervention. Polyphenol supplementation for at least 7 days has a clear moderate benefit on performance in healthy individuals. The performance benefits caused by quercetin supplementation are higher than those of other polyphenols. Further research is needed to confirm the optimal dose, even if a major intake could improve performance response.

Overall, the pooled results show that quercetin is viable supplement to improve performance in healthy individuals.

4.2 Catechins: green tea extract

Although quercetin is the most studied flavonoid in relation to exercise, other molecules are under investigation for their ability to prevent exercise-induced muscle damage and to affect physical performance. As of today, many studies on polyphenols and physical exercise concerned in supplementation with antioxidants like the green tea extract (GTE) from *Camellia sinensis*. GTE extract is rich in polyphenols, with flavonoid structure, including epigallocatechin gallate, epicatechin, epigallocatechin, and epicatechin gallate, which result in a powerful antioxidant activity [22–25]. Green tea supplementation has been advocated as a strategy to improve exercise recovery due to the activity of its catechins with high antioxidant and anti-inflammatory potential [46–48].

Although most studies on green tea have been performed in animals, a considerable amount of data is now available in humans. A green tea extract rich in catechins and caffeine increases the daily energy expenditure in humans. More recently, an acute dose of green tea extract has been evaluated on healthy untrained men in a 30 min cycling test at 60% VO_2 max [46–48].

Other studies showed that GTE supplementation might reduce oxidative stress and promote improvement in the maximal oxygen uptake during cycling to exhaustion. Furthermore, GTE can reduce muscle soreness resultant of eccentric exercise and decrease markers of muscle damage after eccentric exercise, intense aerobic exercise, and strength exercises. Similar effects were not found when a single-dose of GTE was intake before intense muscle-endurance. The effects described for GTE supplementation suggest that GTE supplementation could be a valuable strategy for preserving performance during repeated periods of exercise that cause cumulative fatigue [47, 48].

Jowko et al. have tested the activities of green tea catechins in healthy individuals and soccer players and have been found to be very modest protection from oxidative damage in the first and no effect in the second [47].

Furthermore, it should be understood whether the supply of catechins increases or decreases the performance, in addition to the alleged cellular antioxidant activities. It has been tested a combination of epigallocatechin-gallate and N-acetylcysteine in healthy volunteers who performed eccentric exercise bouts [49]. In another study, the 4-week green tea extract supplement in previously untrained men increased the total plasma antioxidant potential and prevented oxidative damage. In another study, the protective effect of green tea drinks on oxidative stress and muscle damage parameters was also observed in weight-trained men [47, 48].

GTE supplementation protected against oxidative stress is induced by acute muscular endurance test, as well as against muscular damage induced by the training alone. Similar observations were reported in a study [50] about a group of resistance-trained men. In both cited studies, significant decrease in post-exercise plasma creatine kinase activity was noted as a result of supplementation.

However, GTE supplementation provided no protection from exercise-induced muscle damage. On the other hand, a number of previous studies revealed intensified muscle damage and hindered recovery as a result of antioxidant supplementation [47, 48].

Supplementation with GTE prevents oxidative stress induced by high-intensity repeated sprint test in male sprinters. On the other hand, neither protection from exercise-induced muscle damage, nor an improvement in sprint performance was noted after GTE intake. The use of GTE as a supplement is probably not useful in the case of sprinters, at least during the preparatory phase of their annual training cycle. Instead, the effects of taking GTE during the competition phase of the annual training cycle, being associated with a considerably greater exercise load, should be the subject of other research. Supplementation with green tea extract prevents oxidative stress induced by two repeated cycle sprint tests in sprinters. Furthermore, GTE supplementation does not seem to hinder training adaptation in antioxidant enzyme system. On the other hand, neither prevention of exercise-induced muscle damage, nor an improvement in sprint performance is noted after GTE administration [46].

Taken together, data from available studies seem to suggest that catechins can improve physical performance particularly in term of endurance capacity and VO_2 max in untrained subjects, and the same results could be reached in physically active people and well-trained athletes.

In conclusion, it is possible to state that supplementation of green tea extracts before a cumulative fatigue event minimizes muscle damage and oxidative stress in trained athletes. It also has positive effects on neuromuscular parameters related to muscle activation and muscle fatigue. Therefore, the use of GTE as a supplement can be considered a valid strategy in the context of competitive sport of resistance, which aims at the performance of athletes and the recovery of the exercises [48].

4.3 Resveratrol

At the beginning of the 90s, the idea was born that resveratrol, a compound present in red wine, could contribute in part to the "French paradox," the presumed phenomenon for which in France, despite the relatively high consumption of foods rich in acids saturated fat, the incidence of mortality from cardiovascular disease was relatively low, lower than other dietetically comparable countries [51]. Resveratrol (3,4′,5-trihydroxystilbene, RES) is a small polyphenol compound freely available in food supplements, and it is found in various berries, nuts, in the seeds and skins of grapes, red wine, mulberries, peanuts and rhubarb and other plants sources, including traditional Asian medicines [22, 23].

RES is an important activator of the sirtuin proteins and genes (SIRT, silent information regulators), causing an increase in the use of energy, and therefore, reinforcing the mitochondrial function. Sirtuins are silent, but significant regulators of metabolism, cancer, aging, and longevity; they are enzymes associated with the signal transduction pathways connected to stress [52].

Only few studies have investigated resveratrol ability in humans to modulate exercise performance, and some evidence suggests that it could play a role improving endurance capacity. There is a growing interest in the association between RES and exercise, because it has been hypothesized that the administration of RES can produce favorable effects on the rejuvenation of the liver cells, preserves the liver glycogen stores decreased during physical activity, and exercises a regulatory effect on glucose metabolism. To date, most of the studies that have investigated the effect of resveratrol administration on patient outcomes have been limited by their sample sizes.

In a study involving 14 athletes, RES supplementation was shown to inhibit the lipid peroxidation caused by exercise. In a study, it has been demonstrated that a combination of resveratrol and exercise training increased time to exhaustion compared to exercise training. The authors suggested that resveratrol optimizes fatty acid metabolism, which may contribute to the increased contractile force response of skeletal muscles [53]. Despite the inconsistency among reports regarding the topic, it has been suggested that RES delays fatigue by hindering lipid peroxidation, and recently there has been an interest in the capability of resveratrol to modulate physical performance and prevent oxidative stress. Currently, most clinical trials have been conducted with small samples, a wide range of dose levels and groups studied. As a result, it is difficult to establish a specific safety/efficacy range for the RES assay. Many conflicting discrepancies and information must be resolved before recommending the use of resveratrol as a supplement in sports performance [54].

4.4 Polyphenols mixtures

In recent years, the research has focused on studying not only the action of individual polyphenols but also the biological effects of polyphenols mixtures.

In muscular myotubes incubated with polyphenolic extracts of blueberry fruits (*Vaccinium corymbosum* cv. Reka), a dose-dependent protective effect on oxidative stress was observed [55].

The dark chocolate polyphenols were held responsible for some positive effects of dark chocolate consumption during the year. The effects of regular dark chocolate consumption (80 g/day for 2 weeks), rich in cocoa polyphenols, were analyzed on a sample of 20 active men. Plasma metabolites, hormones, and oxidative stress markers were evaluated after prolonged exercise. It has been observed that dark chocolate intake is associated with reduction of oxidative stress markers and increased mobilization of free fatty acids after exercise, but has no observed effect on exercise performance [56].

It has been found that Ecklonia cava (a species of brown alga present in the ocean of Japan and Korea) polyphenols acute preexercise supplementation induces a slight but significant increase in time to exhaustion in healthy human subjects [57].

Anthocyanins represent a class of polyphenols whose use is spreading among sportsmen. They are easy to find in berries and other colorful fruits and vegetables. They can act as antioxidants and anti-inflammatory and therefore can improve recovery from exercise. *In vitro* observations showed anthocyanin-induced activation and endothelial nitric hormone metabolite metabolism and human vascular cell migration. The mechanisms by which anthocyanin intake can improve exercise performance may include effects on metabolic pathways, blood flow, and peripheral muscle fatigue, or a combination of all three. However, in general, the effects of these polyphenols on physical performance are less clear. For example, the use of black currant showed effects on the performance of the exercise; less noticeable effects have instead been analyzed after a cherry intake. Therefore, probably, the benefits could be due to specific food-dependent anthocyanins [12, 14].

Yerba Mate (YM) is a South American plant, rich in polyphenols, saponins, and xanthines, of growing scientific interest because of its metabolic effects. YM has been shown to increase fat utilization during exercise in untrained humans. Its metabolic and physical performance effects were characterized in 11 well-trained male cyclists. YM increased fat utilization during submaximal exercise and improved time trial performance [58].

Montmorency cherry concentrate is used as a supplement by the athletes of the Australian Institute, because it is rich in anthocyanidins. These cherries possess high anti-inflammatory and antioxidant capabilities, can improve sleep and reduce muscle damage and post-exercise pain [59].

Exercise-based studies evaluated the effects of cherry juice supplementation on recovery from maximum strength or resistance (duration>60 min), demonstrating the attenuation of markers related to both inflammation and oxidative stress [60].

Any response linked to accelerated recovery would appear beneficial when considering the large training load experienced by high performance athletes. In reverse, cherry juice (CJ) supplementation had no significant effect on the recovery of Water Polo specific athletic performance. Probably, CJ supplementation may not be necessary for water-based nonweight bearing intermittent sports such as Water Polo [61].

Fourteen male students drank 12 fl oz. of a cherry juice blend or a placebo twice a day for eight consecutive days. A bout of eccentric elbow flexion contractions was performed on the fourth day of supplementation. Isometric elbow flexion strength, pain, muscle tenderness, and relaxed elbow angle were recorded before and for 4 days after the eccentric exercise. Strength loss and pain were significantly less in the cherry juice trial versus placebo. These results show efficacy of the cherry juice in decreasing some of the symptoms of exercise-induced muscle damage [62]. Another study has demonstrated that Montmorency cherry juice consumption improved the recovery of isometric muscle strength after intensive exercise [63].

Regardless, future research should examine the use of CJ in other team sports before CJ can be recommended or excluded as an integrator to improve recovery after sport performance.

Blackcurrant (Ribes nigrum) fruits are a real mine of polyphenols, in fact they are rich in anthocyanins delphinidin-3-rutinoside, delphinidine-3-glucoside, cyanidin-3-rutinoside, and cyanidin-3-glucoside. The health benefits are thought to be mediated by the effect of anthocyanins on inflammatory responses, antioxidant activity, and endothelial function [64]. Moreover, blackcurrant intake increases forearm blood flow at rest, potentially mediated by anthocyanin-induced vasodilation and vaso-relaxation which may affect substrate delivery and exercise performance. It is important to emphasize that the blackcurrants properties are common to all berries (raspberry, blueberry, blackberry, currants, and gooseberries).

Recent studies have revealed a potential ergogenic effect of New Zealand blackcurrant (NZBC) extract intake on physiological and metabolic exercise responses and performance outcomes. In one study, the effect of New Zealand's blackcurrant extract on performance during anaerobic sprint test running in youthful and recreational male soccer players was evaluated. A clear benefit of NZBC's short-term intake on fat oxidation and physical performance has been demonstrated, and the extract seems to benefit the repeated sprint performance only in trained players [65]. Moreover, 7 day NZBC intake augments fat oxidation during 120 min moderate-intensity exercise in endurance-trained females [66].

Another polyphenols-rich fruit is pomegranate; in fact, pomegranate juice (POMj) is rich in flavonols, flavonoids, gallic acid, ellagic acid, quercetin, and ellagitannins, with numerous health benefits during stressful situations [67].

Its antioxidant potential has proven to be superior even to green tea and red wine. According to recent studies, in fact, the pomegranate reduces oxidative stress of macrophages, free radicals, lipid peroxidation, and oxidation of low-density lipoproteins; the inflammatory processes seem to be blocked by the action of ellagitannins. Pomegranate juice is an excellent post-workout because the antioxidants present in the juice of the arils help the muscles to restore their functionality facilitating the supercompensation of exercise; it has a significant impact on acute post-exercise lipid peroxidation and on enzymatic and nonenzymatic antioxidant responses.

Pomegranate extract has been suggested as an ergogenic aid due to its rich concentration of polyphenols, which are proposed to enhance nitric oxide bioavailability, thereby improving the efficiency of oxygen usage, and consequently, endurance exercise performance. Supplementation with pomegranate juice has the potential to attenuate oxidative stress by enhancing antioxidant responses assessed acutely and up to 48 h following an intensive weightlifting training session [68, 69].

The polyphenol *curcumin*, derived from the rhizome *Curcuma longa* L., is a natural antioxidant that exhibits various pharmacological activities and therapeutic properties and has been used to treat a variety of inflammatory conditions and chronic diseases. It has been demonstrated that curcumin can reduce the accumulation of advanced glycation end-products *in vitro* and in animal models, suggesting that this anti-glycation mechanism may relate to the antioxidant effect of the compound. It has been suggested a positive effect of curcumin and Boswellia serrata gum resin supplementation for 3 months on glycoxidation and lipid peroxidation in athletes chronically exercising intensively and further studies will test whether treatment with curcumin can result in a reduction of the accumulation of advanced glycation end-products in muscle tissue, possibly improving muscle performance in the long term [70]. It has been demonstrated that consumption of curcumin reduced biological inflammation, but not quadriceps muscle soreness, during recovery after exercise-induced muscle damage. The observed improvements in biological inflammation may translate to faster recovery and improved functional capacity during subsequent exercise sessions [71].

Honey, natural food produced by the nectar of flowers from bees, is widely used for its precious nutritional and therapeutic values that provide phytotherapeutic properties, with powerful antioxidant, anti-inflammatory, and antimicrobial effects. So far, around 300 types of honey have been recognized with different taste, color, and odor according to the different types of nectar harvested by bees. Honey is an ancient *nutraceutical*, which owes its properties to the richness of polyphenols, which vary according to the floral variety from which it derives, but in general it is made up of flavonoids, between 50 and 500 mg/kg, including galangina, quercetin, kaempferol, and luteolin, which represent the bioactive molecules with a strong antioxidant action.

Honey is an energizing substance useful for sportsmen, and it provides up to 17 g of carbohydrates for every spoon consumed and provides the much needed energy, serving as an economic substitute for the enhancers of sporting activities available on the market. A beneficial effect of honey has been shown in athletes, where if a moderate and regular exercise is able to counteract oxidative stress [20]. In one study, 32 healthy volunteers underwent a short but intense exercise on the ergometer. A significant decrease in serum malondialdehyde levels was observed in subjects who had consumed honey before making a physical effort, with a greater difference for those volunteers who had used it for 3 weeks.

In another study, the effects of honey in 39 road cyclists were examined. In the group that received honey supplementation (70 g), the increase in oxidative stress markers was much lower than placebo, and the antioxidant levels were significantly higher. Ahmad and others examined the effect of different doses of Tualang honey in 20 athletes involved in different competitive sports. The results showed that there was no significant difference between the two different doses and that the maximum antioxidant capacity was observed in both cases 2 h after the honey intake [20].

5. Polyphenols supplementation in exercise: limits and considerations

The use of polyphenols has been designed to improve performance by increasing mitochondrial biogenesis in two ways: polyphenols stimulate stress-related cell signaling pathways that increase the expression of genes encoding cytoprotective proteins such as nuclear respiratory factor; the selected polyphenols (i.e., resveratrol, curcumin, and quercetin) have been reported to modulate muscle function and mitochondrial biogenesis by activating the sirtuins and increasing the activity of the c-receptor co-activator activated by the peroxisome proliferator. Furthermore, some polyphenols improve flow-mediated dilation and endothelial function in humans by increasing the synthesis of endothelial nitric oxide. Polyphenols could help overall athletic performance in sports where the rate of blood flow and maximum cardiac output are important determinants of cardiovascular performance, acting on endothelial function.

Polyphenol supplementation is currently controversial, and at the moment, the use of different exercise protocols, different outcomes, in various physically trained subjects, and the use of a variety of laboratory parameters to demonstrate these effects make it still difficult to assess the effects of polyphenols on physical activity. Therefore, in any case, a detailed description of the type of exercise (e.g., aerobic or anaerobic), the oxidative stress biomarkers used, the characteristics of the subject and the training endpoints examined to allow data interpretation is always necessary.

The evidence is not sufficient to make recommendations for or against the use of polyphenol supplements for recreational, competitive, or elite athletes. Polyphenols have multiple biological effects, and future exercise studies must be studied in an appropriate and specific way to determine the physiological interactions between the exercise and the selected supplement, rather than considering only performance.

Those with higher levels of oxidative stress can clearly benefit more from the antioxidant treatment. An initial screening of the state of oxidative stress is therefore essential. Clearly, individual susceptibility related to the presence of specific genetic variants in key enzymes for ROS detoxification may be another important parameter.

It would be useful to consider the integrated effect of exogenous diet and antioxidant supplementation.

6. Conclusion

The relationship between oxidative stress and sport is really very complex; in fact, the release of free radicals is necessary to stimulate the up-regulation of endogenous antioxidant defenses. In recent years, the consumption of supplements rich in antioxidant compounds by athletes has greatly increased, but a natural intake through the diet is more recommended.

For future research conducted on the performance effects of dietary polyphenols, it should provide adequate detail on the method of blinding and participant follow-up to ascertain whether the study was in fact blinded and report performance data in raw values. These designs would enable researchers to optimize both type and dose of polyphenol supplementation to achieve performance benefit. In addition to the general notes on research reporting, very few studies outlined comprehensive dietary control measures.

Researchers should attempt to quantify the participants' dietary intake of polyphenols, as those with low intakes are likely to respond more favorably to dietary intervention.

Polyphenol supplementation for at least 7 days has a clear moderate benefit on performance in healthy individuals. More research is needed on optimal dose; however, greater intakes could improve the performance response.

The present review summarized the results of studies on the effects of polyphenols intake on exercise-induced oxidative stress obtained in human trials.

The conflicting findings of previous research have brought into question the usefulness of antioxidant supplementation during resistance training. As polyphenolic antioxidants have shown promise as recovery strategies from fatiguing and damaging bouts of exercise, supplementation with polyphenols may be an appealing option to recover from an intense resistance exercise bout. However, it is important to determine whether polyphenol supplementation during a resistance training program will augment or diminish adaptations in muscular strength. Clearly, there is much more to be learned in the exciting field of exercise, oxidative stress, and polyphenols.

Acknowledgements

The author acknowledges the work supported by "Università degli Studi di Napoli *Parthenope*, Bando di ricerca individuale annualità 2017" to Stefania D'Angelo.

Conflict of interest

The author declares that there is no conflict of interest.

Author details

Stefania D'Angelo
Department of Movement Sciences and Wellbeing,
University "Parthenope", Napoli, Italy

*Address all correspondence to: stefania.dangelo@uniparthenope.it

References

[1] Halliwell B, Gutteridge JMC. Free Radicals in Biology and Medicine. Vol. 1999. Oxford: Oxford University Press; 1999

[2] Bast A, Haenen GRMM. Chapter 2: Nutritional antioxidants it is time to categorise. In: Lamprecht M, editor. Antioxidants in Sport Nutrition. Boca Raton, FL: CRC Press/Taylor & Francis; 2015

[3] Reid MB, Haack KE, Franchek KM, Valberg PA, Kobzik L, West MS. Reactive oxygen in skeletal muscle I. Intracellular oxidant kinetics and fatigue in vitro. Journal of Applied Physiology. 1992;**73**:1797-1804. DOI: 10.1152/jappl.1992.73.5.1797

[4] Gomez-Cabrera MC, Domenech E, Romagnoli M, Arduini A, Borras C, Pallardo FV, et al. Oral administration of vitamin C decreases muscle mitochondrial biogenesis and hampers training-induced adaptations in endurance performance. The American Journal of Clinical Nutrition. 2008;**87**:142-149. DOI: 10.1093/ajcn/87.1.142

[5] Webb R, Hughes MG, Thomas AW, Morris K. The ability of exercise-associated oxidative stress to trigger redox-sensitive signalling responses. Antioxidants (Basel). 2017;**6**(3):1-21. DOI: 10.3390/antiox6030063 (Review)

[6] Stear SJ, Burke LM, Castell LM. BJSM reviews: A–Z of nutritional supplements: Dietary supplements, sports nutrition foods and ergogenic aids for health and performance: Part 3. British Journal of Sports Medicine. 2009;**43**:890-892. DOI: 10.1136/bjsm.2009.065417

[7] Kerksick CM, Zuhl M. Chapter 1: Mechanisms of oxidative damage and their impact on contracting muscle. In: Lamprecht M, editor. Antioxidants in Sport Nutrition. Boca Raton, FL: CRC Press/Taylor & Francis; 2015

[8] Sies H. Oxidative stress: A concept in redox biology and medicine. Redox Biology. 2015;**4**:180-183. DOI: 10.1016/j.redox.2015.01.002 (Review)

[9] Kawamura T, Muraoka I. Exercise-induced oxidative stress and the effects of antioxidant intake from a physiological viewpoint. Antioxidants (Basel). 2018;**7**(9):1-19. DOI: 10.3390/antiox7090119

[10] Mankowski RT, Anton SD, Buford TW, Leeuwenburgh C. Dietary Antioxidants as modifiers of physiologic adaptations to exercise. Medicine & Science in Sports & Exercise. 2015;**47**(9):1857-1868. DOI: 10.1249/MSS.0000000000000620

[11] Overdevest E, Wouters JA, Wolfs KHM, van Leeuwen JJM, Possemiers S. Citrus flavonoid supplementation improves exercise performance in trained athletes. Journal of Sports Science and Medicine. 2018;**17**:24-30

[12] Malaguti M, Angeloni C, Hrelia S. Polyphenols in exercise performance and prevention of exercise-induced muscle damage. Oxidative Medicine Celllar Longevity. 2013;**2013**:825928. DOI: 10.1155/2013/825928 (Review)

[13] Sano M, Fukuda K. Activation of mitochondrial biogenesis by hormesis. Circulation Research. 2008;**103**(11):1191-1193. DOI: 10.1161/CIRCRESAHA.108.189092

[14] Myburgh KH. Polyphenol supplementation: Benefits for exercise performance or oxidative stress? Sports Medicine. May 2014;**44**(Suppl 1):S57-S570. DOI: 10.1007/s40279-014-0151-4 (Review)

[15] Wagner KH. Chapter 4: Antioxidants in sport nutrition. All the same effectiveness? In: Lamprecht M, editor. Antioxidants in Sport Nutrition. Boca Raton, FL: CRC Press/Taylor & Francis; 2015

[16] Sellami M, Slimeni O, Pokrywka A, Kuvačić G, D Hayes L, Milic M, et al. Herbal medicine for sports: A review. Journal of the Internaional Society of Sports Nutrition. 2018;**15**:14. DOI: 10.1186/s12970-018-0218-y

[17] Halliwell B, Gutteridge JMC. Free Radicals in Biology & Medicine. 5th ed. New York, NY, USA: Oxford University Press; 2007. ISBN: 9780198717485

[18] Keli SO, Hertog MG, Feskens EJ, Kromhout D. Dietary flavonoids, antioxidant vitamins, and incidence of stroke: The Zutphen study. Archives of internal medicine. 1996;**156**(6):637-642. DOI: 10.1001/archinte.1996.00440060059007

[19] Manach C, Williamson G, Morand C, Scalbert A, Rémésy C. Bioavailability and bioefficacy of polyphenols in humans. I: Review of 97 bioavailability studies. The American Journal of Clinical Nutrition. 2005;**81**:230S-242S. DOI: 10.1093/ajcn/81.1.230S

[20] Cianciosi D, Forbes-Hernández TY, Afrin S, Gasparrini M, Reboredo-Rodriguez P, Manna PP, et al. Phenolic compounds in honey and their associated health benefits: A review. Molecules. 2018;**23**:2322. DOI: 10.3390/molecules23092322 (Review)

[21] Williamson G. The role of polyphenols in modern nutrition. Nutrition Bulletin. 2017;**42**(3):226-235. DOI: 10.1111/nbu.12278 (Review)

[22] Bravo L. Polyphenols: Chemistry, dietary sources, metabolism, and nutritional significance. Nutrition Reviews. 1998;**56**(11):317-333. DOI: 10.1111/j.1753-4887.1998.tb01670.x

[23] Manach C, Scalbert A, Morand C, Rémésy C, Jiménez L. Polyphenols: Food sources and bioavailability. The American Journal of Clinical Nutrition. 2004;**79**(5):727-747. DOI: 10.1093/AJCN/79.5.727

[24] Tzima K, Brunton NP, Rai DK. Qualitative and quantitative analysis of polyphenols in lamiaceae plants–A review. Plants (Basel). 2018;**7**(2):1-30. DOI: 10.3390/plants7020025, 10.3390/plants7020025

[25] Mattera R, Benvenuto M, Giganti MG, Tresoldi I, Pluchinotta FR, Bergante S, et al. Effects of polyphenols on oxidative stress-mediated injury in cardiomyocytes. Nutrients. 2017;**20**(5):1-43. DOI: 10.3390/nu9050523 (Review)

[26] Pérez-Jiménez J, Neveu V, Vos F, Scalbert A. Identification of the 100 richest dietary sources of polyphenols: An application of the phenol-explorer database. European Journal of Clinical Nutrition. 2010;**64**(Suppl 3):S112-S120. DOI: 10.1038/ejcn.2010.221

[27] Miglio C, Chiavaro E, Visconti A, Fogliano V, Pellegrini N. Effects of different cooking methods on nutritional and physicochemical characteristics of selected vegetables. Journal of. Agricultural and Food Chemistry. 2008;**56**:139-147. DOI: 10.1021/jf072304b

[28] D'Angelo S, Cimmino A, Raimo M, Salvatore A, Zappia V, Galletti P. Effect of reddening-ripening on the antioxidant activity of polyphenol extracts from cv. 'Annurca' apple fruits. Journal of Agricultural and Food Chemistry. 2007;**55**(24):9977-9985. DOI: 10.1021/jf071773a

[29] Brglez Mojzer E, Knez Hrnčič M, Škerget M, Knez Ž, Bren U. Polyphenols: Extraction methods, antioxidative action, bioavailability and anticarcinogenic effects. Molecules.

2016;**21**(7):1-38. DOI: 10.3390/molecules21070901

[30] Serafini M, Ghiselli A, Ferroluzzi A. Red wine, tea, and antioxidants. Lancet. 1994;**344**:626. DOI: 10.1016/S0140-6736(94)92017-6

[31] D'Angelo S, Sammartino D. Protective effect of Annurca apple extract against oxidative damage in human erythrocytes. Current Nutrition & Food Science. 2015;**11**(4):248-256. DOI: 10.2174/1573401311666150610210529

[32] D'Angelo S, La Porta R, Napolitano M, Galletti P, Quagliuolo L, Boccellino M. Effect of Annurca apple polyphenols on human HaCaT keratinocytes proliferation. Journal of Medicinal Food. 2012;**15**(11):1024-1031. DOI: 10.1089/jmf.2012.0076

[33] D'Angelo S, Morana A, Salvatore A, Zappia V, Galletti P. Protective effect of polyphenols from Glycyrrhiza glabra against oxidative stress in Caco-2 cells. Journal of Medicinal Food. 2009;**12**(6):1326-1333. DOI: 10.1089/jmf.2008.0285

[34] Pang J, Zhang Z, Zheng TZ, Bassig BA, Mao C, Liu X, et al. Green tea consumption and risk of cardiovascular and ischemic related diseases: A meta-analysis. International Journal of Cardiology. 2016;**202**:967-974. DOI: 10.1016/j.ijcard.2014.12.176

[35] Santos RM, Lima DR. Coffee consumption, obesity and type 2 diabetes: A mini-review. European Journal of Nutrition. 2016;**55**:1345-1358. DOI: 10.1007/s00394-016-1206-0

[36] D'Angelo S, Martino E, Ilisso CP, Bagarolo ML, Porcelli M, Cacciapuoti G. Pro-oxidant and pro-apoptotic activity of polyphenol extract from Annurca apple and its underlying mechanisms in human breast cancer cells. International Journal of Oncology. 2017;**51**:939-948. DOI: 10.3892/ijo.2017.4088

[37] Gomez-Cabrera MC, Salvador-Pascual A, Cabo H, Ferrando B. Viña. Redox modulation of mitochondriogenesis in exercise. Does antioxidant supplementation blunt the benefits of exercise training? Free Radical Biology and Medicine. 2015;**86**:37-46. DOI: 10.1016/j.freeradbiomed.2015.04.006

[38] Visioli F. Chapter 6: Polyphenols in sport. Facts or fads? In: Lamprecht M, editor. Antioxidants in Sport Nutrition. Boca Raton, FL: CRC Press/Taylor & Francis; 2015

[39] Knapik JJ, Steelman RA, Hoedebecke SS, Austin KG, Farina EK, Lieberman HR. Prevalence of dietary supplement use by athletes: Systematic review and meta-analysis. Sports Medicine. 2016;**46**(1):103-123. DOI: 10.1007/s40279-015-0387-7 (Review)

[40] Beck KL, Thomson JS, Swift RJ, von Hurst PR. Role of nutrition in performance enhancement and postexercise recovery. Open Access Journal of Sports Medicine. 2015;**11**(6):259-s67. DOI: 10.2147/OAJSM.S33605. eCollection 2015 (Review)

[41] Somerville V, Bringans C, Braakhuis A. Polyphenols and performance: A systematic review and meta-analysis. Sports Medicine. 2017;**47**(8):1589-1599. DOI: 10.1007/s40279-017-0675-5 (Review)

[42] Belviranli M, Okudan N. Chapter 5: Well-known antioxidants and newcomers in sport nutrition coenzyme Q10, quercetin, resveratrol, pterostilbene, pycnogenol and astaxanthin. In: Lamprecht M, editor. Antioxidants in Sport Nutrition. Boca Raton, FL: CRC Press/Taylor & Francis; 2015

[43] MacRae HS, Mefferd KM. Dietary antioxidant supplementation combined with quercetin improves cycling

time trial performance. International Journal of Sport Nutrition and Exercise Metabolism. 2006;**16**:405-419. DOI: 10.1123/ijsnem.16.4.405

[44] McAnulty R, McAnulty LS, Nieman DC, Quindry JC, Hosick PA, Hudson MH, et al. Chronic quercetin ingestion and exercise-induced oxidative damage and inflammation. Applied Physiology, Nutrition and Metabolism. 2008;**33**: 254-262. DOI: 10.1139/H07-177

[45] Konrad M, Nieman DC. Chapter 10: Evaluation of quercetin as a countermeasure to exercise-induced physiological stress. In: Lamprecht M, editor. Antioxidants in Sport Nutrition. Boca Raton, FL: CRC Press/Taylor & Francis; 2015

[46] Jówko E, Długołęcka B, Makaruk B, Cieśliński I. The effect of green tea extract supplementation on exercise-induced oxidative stress parameters in male sprinters. European Journal of Nutrition. 2015;**54**(5):783-791. DOI: 10.1007/s00394-014-0757-1

[47] Ewa Jówko. Chapter 8: Green tea catechins and sport performance. In Lamprecht M, editor. Antioxidants in Sport Nutrition. 2015. In: Lamprecht M, editor. Antioxidants in Sport Nutrition. Boca Raton, FL: CRC Press/Taylor & Francis; 2015

[48] Machado ÁS, da Silva W, Souza MA, Carpes FP. Green tea extract preserves neuromuscular activation and muscle damage markers in athletes under cumulative fatigue. Frontiers in Physiology. 2018;**17**(9):1137. eCollection 2018. DOI: 10.3389/fphys.2018.01137

[49] Kerksick CM, Roberts MD, Dalbo VJ, Kreider RB, Willoughby DS. Changes in skeletal muscle proteolytic gene expression after prophylactic supplementation of EGCG and NAC and eccentric damage. Food and Chemical Toxicology. 2013;**61**:47-52. DOI: 10.1016/j.fct.2013.01.026 [Epub Jan 31, 2013]

[50] Panza VS, Wazlawik E, Ricardo Schutz G, Comin L, Hecht KC, da Silva EL. Consumption of green tea favorably affects oxidative stress markers in weight-trained men. Nutrition. 2008;**24**:433-442. DOI: 10.1016/j.nut.2008.01.009

[51] Renaud S, de Lorgeril M. Wine, alcohol, platelets, and the French paradox for coronary heart disease. Lancet. 1992;**20**(339):1523-1526. DOI: 10.1016/0140-6736(92)91277-F

[52] Novelle MG, Wahl D, Dieguez C, Bernier M, de Caboa R. Resveratrol supplementation, where are we now and where should we go? Ageing Research Reviews. 2015;**21**:1-15. DOI: 10.1016/j.arr.2015.01.002

[53] Dolinsky VW, Dyck JR. Experimental studies of the molecular pathways regulated by exercise and resveratrol in heart, skeletal muscle and the vasculature. Molecules. 2014;**19**(9):14919-11447. DOI: 10.3390/molecules190914919

[54] Baltaci SB, Mogulkoc R, Baltaci AK. Resveratrol and exercise. Biomedical Report. 2016;**5**(5):525-530. (Review) [Epub Oct 11, 2016]. DOI: 10.3892/br.2016.777

[55] Hurst RD, Wells RW, Hurst SM, McGhie TK, Cooney JM, Jensen DJ. Blueberry fruit polyphenolics suppress oxidative stress-induced skeletal muscle cell damage in vitro. Molecular Nutrition and Food Research. 2010;**54**(3):353-363. DOI: 10.1002/mnfr.200900094

[56] Allgrove J, Farrell E, Gleeson M, Williamson G, Cooper K. Regular dark chocolate consumption's reduction of oxidative stress and increase of free-fatty-acid mobilization in response to prolonged cycling. International

Journal of Sport Nutrition and Exercise Metabolism. 2011;**21**(2):113-123

[57] Oh JK, Shin YO, Yoon JH, Kim SH, Shin HC, Hwang HJ. Effect of supplementation with Ecklonia cava polyphenol on endurance performance of college students. International Journal of Sport Nutrition and Exercise Metabolism. 2010;**20**(1):72-79

[58] Areta JL, Austarheim I, Wangensteen H, Capelli C. Metabolic and performance effects of yerba mate on well-trained cyclists. Medicine & Science in Sports & Exercise. 2018;**50**(4):817-826. DOI: 10.1249/MSS.0000000000001482

[59] Bell PG, McHugh M, Stevenson E, Howatson G. The role of cherries in exercise and health. Scandinavian Journal of Medicine& Science in Sports. 2014;**24**:477-490. DOI: 10.1111/sms.12085

[60] Bell PG, Walshe IH, Davison GW, Stevenson E, Howatson G. Montmorency cherries reduce the oxidative stress and inflammatory responses to repeated days high-intensity stochastic cycling. Nutrients. 2014;**6**:829-843. DOI: 10.3390/nu6020829

[61] McCormick R, Peeling P, Binnie M, Dawson B, Sim M. Effect of tart cherry juice on recovery and next day performance in well-trained water polo players. Journal of the International Society of Sports Nutrition. 2016;**14**(13):41. eCollection 2016

[62] Connolly DAJ, McHugh MP, Padilla-Zakour OI. Efficacy of a tart cherry juice blend in preventing the symptoms of muscle damage. British Journal of Sports Medicine. 2006;**40**:679-683. DOI: 10.1136/bjsm.2005.025429

[63] Bowtell JL, Sumners DP, Dyer A, Fox P, Mileva K. Montmorency cherry juice reduces muscle damage caused by intensive strength exercise. Medicine & Science in Sports & Exercise. 2011;**43**:1544-1551. DOI: 10.1249/MSS.0b013e31820e5adc

[64] Castro-Acosta ML, Smith L, Miller RJ, McCarthy DI, Farrimond JA, Hall WL. Drinks containing anthocyanin-rich blackcurrant extract decrease postprandial blood glucose, insulin and incretin concentrations. The Journal of Nutritional Biochemistry. 2016;**38**:154-161. DOI: 10.1016/j.jnutbio.2016.09.002

[65] Godwin C, Cook MD, Willems MET. Effect of New Zealand blackcurrant extract on performance during the running based anaerobic sprint test in trained youth and recreationally active male football players. Sports (Basel). 2017;**5**(3):1-10. DOI: 10.3390/sports5030069

[66] Strauss JA, Willems MET, Shepherd SO. New Zealand blackcurrant extract enhances fat oxidation during prolonged cycling in endurance-trained females. European Journal of Applied Physiology. 2018;**118**(6):1265-1272. DOI: 10.1007/s00421-018-3858-3

[67] Zarfeshany A, Asgary S, Javanmard SH. Potent health effects of pomegranate. Advanced Biomedical Researche. 2014;**3**:100. DOI: 10.4103/2277-9175.129371

[68] Crum EM, Barnes MJ, Stannard SR. Multiday pomegranate extract supplementation decreases oxygen uptake during submaximal cycling exercise, but cosupplementation with N-acetylcysteine negates the effect. International Journal of Sport Nutrition and Exercise Metabolism. 2018;**28**(6):586-592. DOI: 10.1123/ijsnem.2017-0407

[69] Ammar A, Turki M, Hammouda O, Chtourou H, Trabelsi K7, Bouaziz M, et al. Effects of pomegranate juice supplementation on oxidative stress

biomarkers following weightlifting exercise. Nutrients. 2017;**29**(8):1-13. DOI: 10.3390/nu9080819

[70] Chilelli NC, Ragazzi E, Valentini R, Cosma C, Ferraresso S, Lapolla A, et al. Curcumin and Boswellia serrata modulate the glyco-oxidative status and lipo-oxidation in master athletes. Nutrients. 2016;**8**(11):1-9. DOI: 10.3390/nu8110745

[71] McFarlin BK, Venable AS, Henning AL, Sampson JN, Pennel K, Vingren JL, et al. Reduced inflammatory and muscle damage biomarkers following oral supplementation with bioavailable curcumin. BBA Clinical. 2016;**18**(5): 72-78. DOI: 10.1016/j.bbacli.2016.02.003